Wisconsin on the Air

WISCONSIN ON THE AIR

*100 Years of Public Broadcasting
in the State That Invented It*

JACK MITCHELL

WISCONSIN HISTORICAL SOCIETY PRESS

Published by the Wisconsin Historical Society Press
Publishers since 1855

The Wisconsin Historical Society helps people connect to the past by collecting,
preserving, and sharing stories. Founded in 1846, the Society is one of the nation's finest
historical institutions.

wisconsin**history**.org
Order books by phone toll free: (888) 999-1669
Order books online: shop.wisconsinhistory.org
Join the Wisconsin Historical Society: wisconsinhistory.org/membership

Front cover: WHA microphone detail, WHI IMAGE ID 42114
Page vi: UW-Madison's Radio Hall, ca. 1946, courtesy of the UW-Madison Archives,
#S02765

Printed in Canada
Cover design by Andrew J. Brozyna, AJB Design
Typesetting by Wendy Holdman Design
20 19 18 17 16   1 2 3 4 5

Library of Congress Cataloging-in-Publication Data
Names: Mitchell, Jack W., 1941– author.
Title: Wisconsin on the air : 100 years of public broadcasting in the state that invented it /
    Jack Mitchell.
Description: Madison, WI : Wisconsin Historical Society Press, [2016] |
    Includes bibliographical references and index.
Identifiers: LCCN 2015046195 (print) | LCCN 2016001122 (e-book) |
    ISBN 9780870207617 (hardcover : alk. paper) | ISBN 9780870207624 (e-book)
Subjects: LCSH: Public broadcasting—Wisconsin—History
Classification: LCC HE8689.7.P82 M57 2016 (print) | LCC HE8689.7.P82 (e-book)
    | DDC 384.5409775—dc23
LC record available at http://lccn.loc.gov/2015046195

For my grandchildren, Jonah, Eli, Nina, and Cecily

CONTENTS

Preface   ix

Introduction: The Wisconsin Idea   1

PART 1: EDUCATIONAL
BROADCASTING (1917–1967)   9

  1 Education Learns to Sing   11

  2 The State Stations   25

  3 The State Radio Network   48

  4 A State Television Network?   65

  5 UW Extension   81

PART 2: PUBLIC BROADCASTING
(1967–2016)   97

  6 Public Television   99

  7 And Radio   120

  8 Wisconsin Public Radio   131

  9 Wisconsin Public Television   151

 10 The Ideas Network   174

  Conclusion: Revisiting the Wisconsin Idea   189

  Notes   195
  Index   213
  About the Author   227

PREFACE

Wisconsin on the Air: 100 Years of Public Broadcasting in the State That Invented It* tells the story of what we now call Wisconsin Public Radio and Wisconsin Public Television. The University of Wisconsin put station 9XM, now WHA, "on the air" in 1917. The State of Wisconsin followed a few years later with the station now called WLBL. While WHA may have stretched the truth a bit in calling itself "the oldest station in the nation," it is certainly the oldest and most committed not-for-profit radio station in the nation. Wisconsin invented tax-supported broadcasting in the public interest, while American broadcasting in general sold commercials to generate private profit. *Wisconsin on the Air* explores the people and the ideas that made tax-supported broadcasting happen here.

For forty-nine of its one hundred years, I have been a close observer of public broadcasting in Wisconsin, and from 1976 to 1997 I played a central role as director of Wisconsin Public Radio. As such, biases may lurk in the story I tell, biases that a less involved chronicler might have avoided. For example, I admit to having more interest in radio than in television, and to caring more about public broadcasting's service to democracy than to instruction, although I recognize the importance of both. With those biases, however, I bring insights, I hope, not available to those who have not spent a half century thinking about the purpose of these institutions and how to realize those purposes in practice.

My mentor, Ron Bornstein, says the story of public broadcasting nationally is less about a unified movement than about the clash of individual egos and fiefdoms that popped up in different places over the years. Wisconsin is something of an exception, which may explain why its public broadcasters have maintained national recognition throughout its one hundred years as other players have come and gone. Wisconsin has certainly had its share of large egos and empire builders—Bornstein and myself among them—but unlike other places, Wisconsin has benefited from an underlying philosophy that transcends personalities to give purpose, legitimacy, and continuity to our efforts. That philosophy is the Wisconsin

Idea. It is a philosophy of public service that distinguishes the University of Wisconsin from its peers and the state of Wisconsin from its neighbors. This book begins and ends with the Wisconsin Idea. That idea weaves its way through one hundred years of influential individuals and events that created the Wisconsin Public Radio and Wisconsin Public Television that exist today.

I've been asked why I included television in this story, since 2017 marks the one hundredth anniversary of public *radio*. Wisconsin had no noncommercial television until 1954 and no statewide television network until 1974. I see the two media as intertwined, however, each helping shape the other. Educational television grew directly out of educational radio. It simply added pictures to the philosophy and practice of the older medium. But it was television that drove the national Public Broadcasting Act in 1967, fifty years after 9XM first broadcast voice and music in 1917. That legislation stripped instruction from public broadcasting's national mission and attempted to democratize a service that was seen as elitist. Noncommercial television in Wisconsin embraced the approach with enthusiasm in the late 1960s. Radio took some time to follow television's lead, but then absorbed the democratic approach more thoroughly than the visual medium. In other words, for the first fifty years, educational radio's philosophy shaped television, while in the next fifty, public television's mission shaped radio.

My greatest regret in telling the one-hundred-year story of public radio and television in Wisconsin is all the significant people I do not mention. Most are simply lost to history, their contributions unnoted but undoubtedly important. Others I had to omit to tell a manageable and coherent story. Please regard those I do discuss as representing the thousands I do not. I also tell less of WPT's and WPR's recent history, since only time will test the impact of current innovations. I hope someone will write a sequel to my story in a hundred years or less.

The packrat tendencies of WHA's early leaders are either a blessing or a curse to anyone trying to tell their story. Perhaps they knew they were making history and that every scrap of paper they handled might someday have significance. I confess to not even trying to plow through all of the McCarty and Lighty papers at the Wisconsin Historical Society archives. Rather, I relied most heavily on the manageable, but still voluminous,

WHA station files in the University of Wisconsin Archives. I appreciate the good cheer with which David Null, Cathy Jacob, and other staff carted countless boxes from the archives basement to the fourth floor for my use.

I received unexpected help from Josh Shepperd, a graduate student in Communication Arts at UW–Madison who asked for my counsel on his research into the philosophical and political origins of public service media in this country. I am not sure how much I helped him, but he definitely helped me to tell my story by sharing pertinent material he found in the various archives he explored, especially those of the Rockefeller Foundation, which took a strong interest in educational radio in the 1930s. Josh's frequent visits to my office were always informative, provocative, and enjoyable.

Most of the material in the last half of the book came from interviews with some of the individuals who contributed to or had observed the unfolding story of public broadcasting firsthand, beginning with Karl Schmidt, who started at WHA in 1941 and played a central role thereafter. All of those I interviewed were candid and gracious with their time, including some whose observations I could not include in the book but nonetheless helped shape it. Current staff members of Wisconsin Public Television and Wisconsin Public Radio did their best to cooperate in this endeavor, even though they did not know what I might say about them. Jim Gill of Wisconsin Public Television and Jeffery Potter of Wisconsin Public Radio were particularly helpful in supplying photos.

As always, my wife, Bonnie, proved a tough and skillful editor, tightening my focus and prose. When she finishes a chapter, I feel that it says exactly what I wanted to say, but so much better. As has been true throughout my years at WPR, I sought the reactions to the draft from Jim Fleming, whose values differ enough from mine to keep me honest.

I was pleased that the Wisconsin Historical Society's Kathy Borkowski and Kate Thompson reacted quickly and enthusiastically to my idea for the book and supported the project through its two-year evolution. They assigned Sara Phillips to edit *Wisconsin on the Air*, a very happy choice for me. She made me verify more facts than I might have preferred, clarify points I thought were perfectly clear, and tone down passages where my emotions got out of hand. The book is stronger as a result.

INTRODUCTION

The Wisconsin Idea

O n a dreary December afternoon in 1978, I heard a commotion outside my WHA radio manager's office. "I am looking for Jack Mitchell!" a voice shouted, and a reporter and a photographer from Madison's *Capital Times* newspaper burst through the door. In full Mike Wallace "gotcha" mode, they began barking questions and snapping pictures. My answers and a lot more turned up in a multistory, page-one exposé with the title, "The Selling of WHA: Some Fear Loss of Quality."[1] The second and third stories appeared the next day with the headline, "The Selling of WHA: Is the Wisconsin Idea Dead?"[2]

The stories did include the legitimate news that on January 1, 1979, Madison stations WHA, an AM station, and WERN, an FM station, would begin to have separate program schedules and to provide complementary programming under the generic identity Wisconsin Public Radio. The two stations were presenting virtually identical programming, a mix of music and talk, but as Wisconsin Public Radio, WHA would specialize in news and informational programming while WERN would emphasize classical music and cultural material. Between the two stations, Wisconsin Public Radio would provide more hours of each type of programming, giving listeners choices throughout the day. I viewed the change as purely positive, a win-win for both music and news listeners, but some saw it as sinister. The reporter opined, "The quality of the expanded programming is less important to its architects than is the size of the audience it is supposed to attract."[3]

While the introduction of dual service was the real news, the *Capital Times* articles dissected what I had done since I had taken over as head of WHA two years earlier, in late 1976. The articles quoted an unnamed university official—I suspected the president of the UW system—complaining that he did not want to hear about abortion with his breakfast. Emeritus political science professor David Fellman, a frequent and popular speaker on WHA and WERN, told the reporter, "I think the new people who took over a few years ago regarded the station as stodgy and old-fashioned, and it seems to me more and more they've been aping commercial radio."[4] In a sense, Fellman was right. Public radio was adapting, as commercial radio had, to the way listeners were using the medium in the television age. When television replaced radio as the dominant broadcast medium, radio became a background companion to routine activities such as driving, washing dishes, or brushing teeth. Listening patterns followed the comings and goings of daily life more than a station's program schedule. The media environment had changed, and I was trying to adapt WHA and WERN to those changes. The adaptations I instituted, however, did not signify a change in purpose. Rather than euthanize the Wisconsin Idea, as the newspaper articles suggested, I meant to resuscitate it.

The skeptics were correct to worry, however, for the Wisconsin Idea provided the central rationale for university- and state-supported broadcasting. The radical concept that the university should serve all residents of the state, not just students on campus, animated the campus as educational radio emerged from the University of Wisconsin physics department in 1917. The Wisconsin Idea envisioned the university as an instrument of democratic change. It sought to pluck the university from its ivory tower and have it grapple with the real problems of ordinary people where they lived and worked. University of Wisconsin president Charles Van Hise and his close friend, Governor—and later Senator—"Fighting" Bob La Follette championed the Wisconsin Idea, but it was Professor Charles McCarthy, a scrappy, 127-pound football All-American from Brown University and a son of poor Irish immigrants, who defined it and explained it to the nation in his 1912 book, *The Wisconsin Idea*, with an introduction by Theodore Roosevelt.[5]

The Wisconsin Idea describes a full array of progressive reforms undertaken in the state early in the twentieth century. The Wisconsin Idea, McCarthy wrote, responded to the concentration of wealth and power in the hands of a few at the expense of the middle class. McCarthy explained, "Our civilization, with its wealth and prosperity, must be made to exist for its true purposes—the betterment, the efficiency and the welfare of each individual."[6] Capitalism inevitably produced a small number of big winners who used their wealth and power to cement their privileged position at the expense of everyone else. They exercised economic power over consumers and smaller businesses. They exploited their workers. They exercised political power by buying politicians to do their bidding. They owned judges. McCarthy foresaw the day when a small group of powerful families owned most of the wealth and the rest of the population drifted into poverty. The Wisconsin Idea sought to empower the middle class in a struggle against the rich and powerful. Education and political reform would help individuals prosper and grow to their full potential.

Professor McCarthy headed the Legislative Reference Bureau. Under its auspices and with the help of fellow UW faculty members, he drafted legislation to right the balance of power. Some legislation attacked concentrated economic power through laws regulating business, establishing workers' compensation, protecting the environment, and limiting child labor. It enacted a state income tax. Other laws addressed political corruption by establishing a professional civil service system. Some legislation gave power directly to the people through primary elections, the direct

Professor Charles McCarthy with staff at the Legislative Reference Library. McCarthy published *The Wisconsin Idea* in 1912. IMAGE COURTESY OF THE UW-MADISON ARCHIVES, #S06554

election of US senators, and the ability to recall elected officials. Theodore Roosevelt's introduction described Wisconsin as "literally a laboratory for wise experimental legislation aiming to secure the social and political betterment of the people as a whole."[7]

McCarthy said he sought to empower the individual—"the man," in his phrase. "Why not invest something in the farmer and the mechanic so that he will become more efficient, so that he will have a better home, better prospects, and greater skills?" he asked. "Why not teach him how to live so that he may be strong and vigorous; why not show him his rights under the law?"[8] Education was the answer to these questions, and so the university landed at the center of the Wisconsin Idea. An activist university could help empower common people to thrive in a system McCarthy saw as stacked against them. Shortly after assuming the presidency of the University of Wisconsin in 1903, Van Hise made this philosophy clear: "I shall never be content until the beneficent influence of the University reaches every home in the state."[9]

McCarthy's chapter on education praised the university's work in agricultural extension, which, he said, created the state's dairy industry. Agricultural extension was not unique to Wisconsin—all land-grant institutions engaged in it—but the university in Madison embraced it more vigorously and more effectively than any other institution in the country. The university's unique General Extension division, which McCarthy had helped organize in 1906, had even more potential for good than the agricultural extension because it sought to implement a broader concept of mass education. McCarthy charged the new division with three tasks: facilitating fair and impartial debate on important issues, offering courses by correspondence, and organizing public lectures.[10] The university gave the division its own faculty dedicated to enriching everyone in the state. Ninety-eight professors and other faculty "of the highest rank" went to factories, towns, and villages to teach classes. They supervised more than five thousand students taking correspondence courses.[11] They also promoted grassroots democracy, engaging broad swaths of the public on issues previously dealt with quietly by elites, special interests, and, too often, corrupt politicians. McCarthy reported that General Extension's Department of Debating and Public Discussion lent eighty thousand articles on public issues to individuals and groups throughout the state.[12] He argued that

University president Charles Van Hise, champion of the Wisconsin Idea.
WHI IMAGE ID 33714

General Extension's activities, supported by public money, not private, would ensure free speech in the discussion of public issues.

McCarthy concluded *The Wisconsin Idea* by reiterating the theme of helping people help themselves, which, he emphasized, differs from socialism. Rather than adopt socialist remedies, he urged the nation to follow Wisconsin's example of neutralizing privilege through education along with economic and political reforms. He told the nation to use "hope and encouragement" to "make every man more efficient so that the door of opportunity may always be open before him."[13] He asked businesspeople to act in their enlightened self-interest by supporting state investment in "hope, health, happiness, and justice."[14] The Wisconsin Idea was a model for progressive people everywhere.

While *The Wisconsin Idea* included chapters on education, political reforms, control of business abuses, and social justice, the book did not touch another theme in the progressive reform movement, the role of print media such as magazines and newspapers. Media could reinforce the power of wealth, or they could support the more democratic values of the reformers. Yellow journalist William Randolph Hearst and his fellow newspaper barons across the country were as much members of the wealthy elite as the industrialists, bankers, or railroad tycoons. The application of steam power to the printing press and the concentration of the population into large cities provided newspapers the opportunity to build circulation and profit. To make the most of that opportunity, newspapers needed to attract mass audiences who responded to topics less elevated—less serious—than the academically based reformers thought they should. Sex, crime, and violence attracted audiences to newspapers in the early twentieth century just as reliably as they draw audiences to mass entertainment today. The progressives saw two interconnected problems with the press of the day. Newspapers built readership with content that did nothing to create an informed and involved citizenry necessary for a real democracy. Their owners then used their power to protect their own privileged positions and those of their advertisers.

In the summer of 1912, the same year Charles McCarthy published *The Wisconsin Idea*, University of Wisconsin president Van Hise invited a hundred newspaper editors and critics to the Madison campus. He asked them to address the problem of "Commercialism and Journalism."[15] For three days they described the problems that profit-driven newspapers created and their possible solutions. All proposed solutions centered on non-commercial, not-for-profit newspapers operated by local governments, foundations, or public universities. Universities were deemed particularly appropriate since they were already in the business of research (investigation), teaching (reporting), and public service, and were protected under the umbrella of academic freedom. Hamilton Holt, editor of the *Independent*, a progressive magazine, had campaigned for an alternative to commercial journalism for a decade and brought his message to the Madison conference. "The ordinary commercial press," he told the delegates, could not provide the information needed for citizens to make informed opinions on the issues of the day. Nor, he said, could such papers "provide competent discussion of pending issues from different points of view."[16] He concluded that commercial journalism could never fill these vital functions for democracy because "it does not PAY to be as thorough and impartial as the ideal paper should be. A self-supporting journal must be sensational. It must give undue prominence to spectacular events and crowd out quieter but more important moments."[17]

On an August day in 1912, Holt's oratory reverberated in the auditorium of Music Hall, a small chapel-like building on the south side of the green lawn that sweeps up Bascom Hill in the center of the UW campus. Across from Music Hall on the north side of that lawn stands Science Hall, a hulking Romanesque Revival, where physics professor Earle Terry had toiled since 1910 with his graduates students to build a functioning wireless transmitting device. It took five more years after the newspaper conference and the publication of *The Wisconsin Idea* for Professor Terry and his students to demonstrate the possibilities of "broadcasting," a technology that had the potential to make real the hopes of McCarthy, Van Hise, and Holt.

By the time Terry's radio station broadcast on a reasonably regular schedule, however, the Progressive movement of Theodore Roosevelt, Woodrow Wilson, and Bob La Follette had lost its luster nationally and dimmed even in the home state of the Wisconsin Idea. The horrors of World War I drained the Progressive movement of its optimistic vision of a more rational and democratic society. The Progressive political agenda of McCarthy and La Follette gave way to the business-focused "Roaring Twenties." President Calvin Coolidge spoke for the majority of Americans when he told a convention of newspaper editors in 1925, "The chief business of the American people is business."[18] And business would grab the new technology of radio for its own profit-oriented purposes.

An attenuated Wisconsin Idea continued, however, in the University of Wisconsin, and even in Wisconsin state government. University experts stood ready to assist state government in crafting legislation as McCarthy had done, but state leaders did not necessarily seek or accept the help as readily as they had before the Great War. The university continued to help individuals improve their skills, widen their horizons, advance their careers, and participate more effectively in a democratic political system, but without reference to fighting entrenched power. This newly refined Wisconsin Idea claimed "the boundaries of the university are the boundaries of the state."[19] This concept shaped the thinking of radio pioneers Professors Earle Terry, William Lighty, Andrew Hopkins, and Henry Ewbank in the 1920s. As the first electronic mass medium, radio provided a perfect vehicle to carry the university to the boundaries of the state. It offered a liberal education (Lighty), provided useful information

(Hopkins), and fostered democracy (Ewbank). As America turned broadcasting over to commercial private enterprise, Wisconsin created a unique tax-supported broadcasting system dedicated to the public interest rather than to private profit.

Wisconsin on the Air explains how, over one hundred years, the goals of these progressive thinkers grew into today's Wisconsin Public Radio and Wisconsin Public Television. There were certainly growing pains. As director of Wisconsin Public Radio for more than twenty years, I experienced some in the 1970s, 1980s, and 1990s, but those I experienced were not as severe as those that other leaders of noncommercial radio and television endured as they sought to nurture their alternative to profit-seeking media in Wisconsin. Through one hundred years, these leaders adapted their broadcasting to a changing environment, but they never lost the genetic imprint of the Wisconsin Idea.

1

EDUCATION LEARNS TO SING

1917–1929

N o one knows the exact date, but it happened during the first three months of 1917. Physics department assistant professor Earle Terry and his wife, Sadie, invited a group of faculty, deans, and friends to their home to hear the "first broadcast" of the University of Wisconsin radio station. For several years, Professor Terry and his students had been transmitting the dots and dashes of Morse code. Only the geeks of the day who shared their interest in radio technology could decipher it, but this night would open the transmission to all who listened. In Terry's Science Hall lab, graduate student Malcolm Hanson had rigged a telephone mouthpiece to capture the sound from the horn of a phonograph.[1] When his guests gathered, Terry called Hanson and said, "We are all ready." Hanson flipped the necessary switches to excite the wire strung between the top of Science Hall and the chimney of the old university heating plant behind it. He placed the phonograph needle on a record, and a receiver in Terry's living room began to emit the faint sound of a piano playing "Narcissus," a popular tune of the day.[2] The guests were underwhelmed. Years later, one of them said that she liked to think that all the guests "were as dumb about the whole thing as I was."[3] Not a single guest realized that one hundred years of broadcasting from the University of Wisconsin had just begun.

The lack of enthusiasm in the Terry living room that night matched the lack of enthusiasm in the professor's academic home, the physics department. Pioneer broadcaster Edgar "Pop" Gordon dubbed the department's attitude "scornful."[4] The physics faculty prided themselves on

Science Hall, where Earle Terry and his students built 9XM. The Chemical Laboratory (600 N. Park) to the right provided the first home for WHA-TV in 1954. WHI IMAGE ID 58314

the theoretical nature of their work and looked down on the "engineers," who sought practical applications for their theories. Using radio gear to reach a broad audience had little to do with the physics of radio waves, and Terry's colleagues punished his devotion to this peripheral activity. They objected to the time he wasted on radio and to the noise his radio transmitter poured into the offices, labs, and classrooms of Science Hall and later Sterling Hall. Inflicting the ultimate academic slight, the physics department did not promote him to associate professor until he had served almost twenty years as an assistant professor. As one graduate student observed, the physics department tolerated but never supported him.[5]

The only other professor who shared Terry's interest in radio and with whom he collaborated was, in fact, an engineer. In 1914, engineering professor Edward Bennett received a license from the federal government for experimental station 9XM, but he soon turned the project over to Terry, whose passion for the enterprise exceeded his own. Explained Bennett, "Professor Terry's vision extended beyond the stage of experimentation with physical principles and properties. When telephonic transmission [voice and music rather than dots and dashes] became a possibility, Professor Terry grasped the significance of radio broadcasting for the Extension work of the University."[6] The Wisconsin Idea, extending the boundaries of the university to the boundaries of the state, drove Terry to push beyond Morse code and make broadcasts accessible to anyone with a receiver.

Transmission of voice and music required glass vacuum tubes, which were not yet manufactured commercially. The only way to get them was to blow molten glass into the shape needed, install the required electronics, and then remove the air from the tube to create a vacuum. The process presented many opportunities for failure, but Terry mastered the techniques and taught them to his students. After two years of experimenting, they produced tubes that worked well enough to transmit the feeble sound that the guests in Terry's living room could barely hear that winter night in 1917.[7] Because tubes blew up frequently and sometimes dramatically, students needed to continuously produce replacements. They became so skillful that they supplied tubes to other fledgling broadcasters.

Partisans of the University of Wisconsin's pioneering efforts in radio have proclaimed 9XM (later WHA) "the oldest station in the nation." This boast headlines the historical marker affixed to Vilas Hall, which now houses public broadcasting on campus. In reality, no broadcast pioneer can identify precisely when a station began. Each station developed

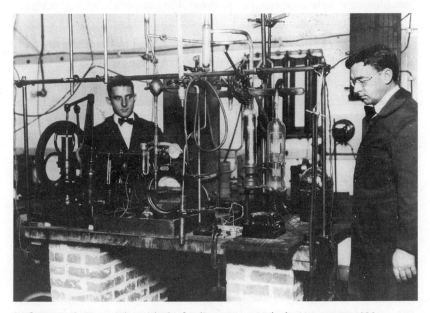

Professor Earle Terry, right, with chief radio operator Malcolm Hanson, ca. 1920.
IMAGE COURTESY OF THE UW-MADISON ARCHIVES, #S06927

incrementally, starting with point-to-point messages in Morse code, eventually adding voice and music, inviting people to listen to these "broadcasts," and, ultimately, producing a full schedule of broadcasts aimed at a broad audience. Wisconsin followed that progression. It was not until four years after 1917's "first broadcast" that 9XM announced a limited broadcast schedule. It took ten more years to produce a full and reliable schedule.

Relicensed as WHA in 1922, 9XM's significance in broadcast history has less to do with being first to broadcast than with being first to implement a public service philosophy of broadcasting. The two great progressive leaders, Professor Charles McCarthy and President Charles Van Hise, died before WHA replaced 9XM, but their Wisconsin Idea dominated the campus when Professor Terry started his work in 1910. Terry embraced the ideal of the university as an empowering force for all state residents. Unlike the engineers, physicists, and tinkerers who developed radio technology at other universities early in the twentieth century, Terry cared about the content his transmitter would broadcast. Terry understood that the real potential for radio was less about point-to-point communication (such as ship to shore) and more about reaching many people simultaneously, in "broadcasting." He conceived of radio as a mass medium with receivers "more common than bathtubs in Wisconsin homes."[8] Terry envisioned radio's power to do good, to educate, to inform, and to inspire large numbers of people simultaneously. The first program director at 9XM, Professor William Lighty, wrote that Terry and Hanson, "who approached the problems . . . from the technical side, also had the uncommon insight into its social possibilities."[9] Others gave similar descriptions of Terry's social vision. Only such vision explains the many hours he and his graduate students devoted to the radio station. They far exceeded their commitment to the physics department and received little acknowledgment or reward for their efforts.

Terry accepted the appointment as the first manager of 9XM, a part-time assignment, in addition to his other professorial duties. With no budgeted staff, he needed to enlist the services of other faculty and graduate students to operate and program the station. He found coconspirators in Andrew "Andy" Hopkins of the Agriculture College and William Lighty, head of correspondence study for General Extension. Both came to the university from adult education backgrounds. They had worked to "Americanize" the wave of immigrants who flooded cities early in the twentieth century. Ensconced at the University of Wisconsin, each would continue his commitment to adult education by using radio.

Hopkins had performed his adult education work for the YMCA in Milwaukee, where he taught English and vocational skills to the city's mostly German immigrants.[10] Hopkins moved to Madison when the College of Agriculture hired him to disseminate to the general public information about advancements in agriculture and home economics. Before radio broadcasting emerged as a useful technology, he "broadcast" useful information via available print media. When the Agriculture College created a Department of Agricultural Journalism to take over this function and to train students for media careers serving rural audiences, Hopkins became its chair and visionary leader. Like his peers in some other midwestern universities, Hopkins saw radio as a better way to do what he was already doing in print. It was a perfect means of reaching the 190,000 small (seventy-acre or less) farms scattered throughout the state, particularly when the literacy rate among state farmers was only 60 percent.[11] Even those who could not read English could understand it when they heard it on the radio. The station's earliest transmissions in Morse code, and then voice, consisted of weather reports around the noon hour so that farmers could listen during their midday dinners. Market reports from the state department of agriculture followed the weather. Adding news from Hopkins's agricultural journalism department created something like a noontime "farm" program.

While Hopkins served on the university's radio committee for decades and his agricultural journalism department made the largest commitment to programming for 9XM/WHA, General Extension's Professor Lighty threw more of his heart and soul into educational radio than anyone other than Terry himself. Indeed, Lighty resembled Terry. Each was an inner-directed individual, true to his own vision, indifferent to what others thought. While beloved by many, neither man fit the conventional academic model nor played academic politics very well. Lighty had spent a decade working for the St. Louis Ethical Movement, which educated and organized workingmen. He said he sought to motivate working people who had been "uninterested in exercising—or lacked the skills to exercise—influence to address their own problems."[12] His ten-year commitment to that challenging task in St. Louis drove him to a state of "nervous exhaustion." Lighty and his family fled St. Louis and retreated to peaceful northern Wisconsin. Charles McCarthy found him there in 1906 and recruited him to help organize President Van Hise's new General Extension division.

Lighty accepted McCarthy's invitation, although he worried that McCarthy's vision of empowerment tended toward practical and vocational skills while he favored history, literature, and cultural enrichment. Lighty cited the Chautauqua circuit as his model.[13] The Chautauqua tent shows traveled among rural communities, bringing musical and dramatic presentations and lectures that sought to inform, entertain, and inspire. They brought to rural residents some of the cultural opportunities available to those who lived in urban areas. Radio, wrote Lighty, should function as a primary socializing agent that would aid in "the rationalizing of all citizens."[14] He did not support broadcasting full courses on the radio. Rather, he saw radio's positive contribution to society more broadly, "to secure broadcasts that have a general human interest appeal for the vast invisible audience, and at the same time to interpret the true spirit, the life and the work of the university, as well as to instruct, stimulate, and enrich the lives of listeners."[15]

Lighty's General Extension colleague, Lilia Bascom, said he was interested in bringing cultural programs to the people of the state, "for example, good music, not jazz; talks on political questions; lectures and university events."[16] Good music meant classical music to Lighty. A very traditional man, he sought to raise the cultural standards of those less fortunate in their tastes than he. Not surprisingly, his attempt to elevate tastes was not always appreciated by those whose tastes he targeted for elevating. In 1925, Professor Terry received a request from listener C. H. Alzmeyer: "Give me something with a melody and you will git [sic] the applause." He suggested songs such as "Carry Me Back to Old Verginia [sic]." He wanted fiddle tunes. "Fiddle don't mean a VIOLIN," he clarified.[17] Terry's response might have been written by Lighty. "Having been brought up on a farm myself, I think I understand quite well the character of the programs you would most enjoy." But WHA, he said, broadcast only material of merit. "The air is overcrowded every night with jazz and other worthless material, and it would be quite beneath the dignity of the university to add to it."[18]

He closed with the emphatic promise that WHA would never broadcast "old time fiddle music."

The radio station benefited from Professor Lighty's unrestrained, and generally uncritical, enthusiasm for all forms of adult education. According to his colleague Andy Hopkins, Lighty was never critical of others, and no one was ever critical of him. Amused perhaps, but not critical: "He incited the best in others."[19] While his colleagues may have snickered at his formality and eccentricities, such as riding a horse to work as they passed him in their automobiles, they seem to have liked and respected him. Yes, he was neurotic, but no one denied his idealism. Hopkins described him as a very kind man: "He loved people. . . . He wanted to improve society. I think it was his impelling motive in all his educational work."[20]

Lighty tried to entice as many faculty members as possible from all university departments to prepare "talks" for the radio station. He asked faculty experts to write the talks, just as they might write short articles for publication in print. They could read the talks themselves or Lighty could read them on the air. Recollections differ on Lighty's success in attracting the biggest names on campus. Lighty boasted of one big catch after another. In 1924, he bragged that three hundred faculty or staff members had appeared on the station that year.[21] Others said most faculty regarded radio as a plaything "beneath the dignity of a true educator."[22] The chairman of the economics department put his department's refusal diplomatically, saying, "There seems to be an oversupply of modesty, though I urged them all to volunteer."[23] Perhaps expressing such modesty, the assistant head of the university library did not find the library particularly interesting: "This leads me to feel that a twelve-minute broadcast on the 'History of the University Library' would be a flop. Lord knows there has been too much broadcasting on uninteresting subjects and I do not care to add to the number."[24] A few years later, an economics professor declined because he used charts in his presentations, saying, "In their absence, I fear that my series will be rather sketchy."[25]

Perhaps faculty feared having the same experience as Charles E. Brown of the State Historical Society. He reported delivering his talk in a "telephone booth" draped with heavy curtains. The booth, he said, had "no air, no sound, and no hope for him who entered there."[26] He described lights that flashed directives such as "begin," "faster," "slower," "one minute," and "end." The lights so distracted him from his script that he was unsure whether he had read intelligently or "mumbled it to an unseen friend or foe." He concluded, "Weak with nervous exhaustion and heavy perspiration, I stepped limply from the ordeal, hoping that never again would I be called upon to participate in this strange new field of broadcasting."[27]

At a faculty meeting, the chair of Professor Terry's own physics department opposed faculty participation because radio had "no experimental significance."[28] Another scientist responded that radio "had significance as a social experiment."[29] University president Edward Birge provided no support to Lighty's efforts. In fact, the president insisted on receiving copies of all faculty talks five days before broadcast to establish, he said, a written record in case he received complaints. There is no indication that the president censored the talks submitted to his office, but he approached the radio experiment more with caution than enthusiasm.

Other than agricultural journalism, only one academic unit provided consistent support for Lighty's programming. His home unit, the General Extension division, paid part of the salary of a faculty member in the Department of Music, and Lighty tapped that resource for the radio station and created its first "star." Edgar "Pop" Gordon provided talks before live music performances early in the development of 9XM, including a concert by Pablo Casals. In 1922, he conducted an evening music appreciation course ungracefully titled, "The University Extension Instruction Service by Radio Broadcast on the Appreciation of Music."[30] He talked about music, played music, and sometimes asked listeners to sing along. Wrote one listener, "Put me down as one of the participants in the Radio Chorus singing *America* last night. As I was alone in the house, my intention at first was to merely listen in, however, after it got underway, I could not resist rising on my feet and singing."[31]

Although live broadcasts of UW athletic events were not educational by traditional definitions, Lighty enthusiastically scheduled them, particularly basketball games. The same telephone lines installed to broadcast live concerts from the Armory (the Red Gym) allowed live broadcast of sports from that location. Those lines connected the Armory to the WHA broadcast studio in Sterling Hall. WHA had relocated its studio and transmitter to Sterling Hall when the physics department moved there in 1917. Lighty scheduled these sports broadcasts to entice listeners to discover WHA and, he hoped, the other material the station broadcast. The complaints that poured in when technical difficulties interfered with sports broadcasts demonstrated their popularity. Lighty might have been happier if technical problems with other programming had generated a similar level of frustration.

Filling enough broadcast hours became an increasingly important challenge for Lighty. Lilia Bascom remembered him telling reluctant participants that WHA needed to broadcast more than two or three hours a day or the federal government would give the frequency to a commercial broadcaster prepared to provide far more programming—probably including jazz and "fiddle music."[32] Lighty's concern was justified. In its first decade of operation, WHA's programming consisted of the weather forecast, current prices for livestock and other agricultural products, and farm and home economics information for one hour at midday. The station returned to the air some evenings for an hour or two of educational talks, music appreciation, and live broadcasts of concerts and athletic events.

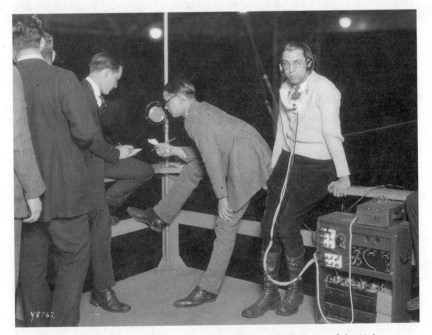

A live broadcast of a basketball game from the Red Gym at the heart of the University of Wisconsin campus in 1922. Live music was also broadcast from this location.
IMAGE COURTESY OF THE UW-MADISON ARCHIVES, #S00024

The station needed to do much more to sustain its spot on the radio dial in the face of a frenzied gold rush for frequencies in the 1920s, when broadcasters realized they could get rich selling advertising around popular entertainment programs. A license to broadcast was virtually a license to print money and few not-for-profit broadcasters withstood the onslaught.

Lighty's challenge in generating programming for WHA paled, however, when compared to the technical and political obstacles faced by Professor Terry. Radio broadcasting in the 1920s resembled the Wild West. Broadcasters took to the air willy-nilly, interfering with each other's signals. Secretary of Commerce Herbert Hoover brought some order to the chaos by requiring licenses for stations seeking to broadcast. The licenses determined the hours a station could operate and its authorized radio frequency. The flood of new stations seeking airtime put pressure on existing stations—including "the oldest station in the nation"—to share the limited number of viable frequencies given the primitive technology of the time. Stations operated by not-for-profit entities with limited resources found themselves pitted against commercial enterprises with broad schedules of entertainment drawing large audiences. The not-for-profits landed at the bottom of the priority list for good broadcast hours and frequencies. Hoover and his staff argued that commercial stations reaching large audiences provided more "public service" than small stations with limited programming and few listeners.

Worse yet, the federal government forced WHA to switch frequencies—its spot on the radio dial—seemingly every few months. Such switching limited a station's ability to build a regular audience. Moreover, most years, WHA closed down for the summer months as well as for frequent technical modifications and upgrades. (The station remained a physics project, after all.) Whatever the value of its content, WHA's service was neither extensive nor reliable. Throughout the 1920s, WHA remained more a work in progress than a reliable public service. Toward the end of the decade, that reality began to change.

In 1925, the University Board of Regents sought a new leader to reinvigorate the progressive activism that had declined after the death of President Van Hise in 1917. They made a surprising choice in Glenn Frank. At only thirty-nine years old, Frank became the youngest university president in Wisconsin history. More surprising, he had no advanced degree nor any

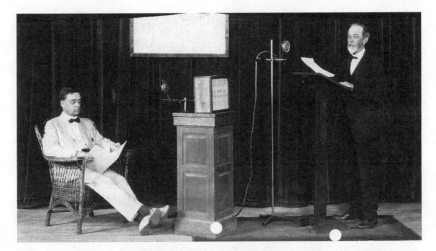

Professors Earle Terry, left, and William Lighty pose in the improvised Sterling Hall studio in 1923. IMAGE COURTESY OF THE UW-MADISON ARCHIVES, #S12746

significant experience as an academic. He was, however, something of a celebrity. He was a writer and public speaker and was sometimes mentioned as a potential president, not of the University of Wisconsin, but of the United States. In addition to his writing and public speaking, Frank edited a nationally prominent progressive magazine. These credentials may have made him a logical choice to lead the Wisconsin Idea of progressive reform and university activism, but they did not make him a good choice to preside over a major research university, especially one where a conservative academic culture belied its public image as a progressive hotbed. Frank never enjoyed the confidence of the faculty. In addition, the La Follette dynasty, which was synonymous with Wisconsin progressivism, did not welcome this interloper and potential rival. Nonetheless, Frank's administration elevated WHA from a physics experiment supported by the uncompensated time of idealistic volunteers to a permanent part of the university's mission.

The handful of advocates for WHA did not perceive Frank as an ally when he arrived on campus. Indeed, early in his administration, Frank, like his predecessor Edward Birge, declined to provide any funding to the station beyond what the physics department, General Extension, and Agricultural College allocated from their own resources. As Professor Hopkins remembered it, "There can be little question about the fact that President

Glenn Frank was lukewarm in his attitude toward the work of Professor Terry and the value of the radio station to the university."[33]

Two years into the Frank presidency, the speech department hired a young assistant professor named Henry "Heiny" Ewbank, whose academic work centered on the importance of debate and public discussion in a democracy and whose major book carried the title *Discussion and Debate: Tools of Democracy.*[34] "A democracy can be permanently successful only if a majority of its citizens have intelligent and considered opinions on matters of public concern," he wrote, and added, "The discussion of public problems is essential to the formulation of this intelligent public opinion."[35] Not surprisingly, Ewbank saw a vital role for radio in promoting his vision of open and rational public debate.

Ewbank joined Terry, Lighty, and Hopkins on the informal radio committee and led them in confronting President Frank with a simple choice: either support WHA or kill it. The committee told Frank the station needed better facilities, money, and "a definite assignment of responsibility to see that satisfactory broadcasting conditions are maintained."[36] As an alternative, they said, the university might close down WHA and try to produce programs for privately owned stations such as WTMJ in Milwaukee. The president chose to support his radio station. He called a meeting of university department heads and asked each to pledge a certain amount of programming from their faculty and staff.

Frank formalized the University Radio Committee, dismissing Lighty as committee chair and station program director, a blow the idealistic educator never fully forgot nor forgave. Frank may have made the decision because Lighty lacked administrative skills, or because his obsession with radio took him away from the job the university paid him to do, directing correspondence study in General Extension. With Lighty pushed aside, President Frank "dumped the squawking, misunderstood foundling on the desk of Henry Ewbank because no one else would take it."[37] Ewbank became a national leader in teaching and research about radio, but he insisted he knew nothing about the medium when he accepted the assignment. "I didn't start from scratch; I started behind scratch," he recalled in 1946.[38] Nevertheless, he remained chair of the Radio Committee and its successor organizations for the next thirty years. As much as any other individual, he shaped educational broadcasting in the state and fostered its unusual emphasis on facilitating democracy.

President Frank may have had reservations about radio when he took office in 1925 but they had changed to enthusiasm by the time he left in 1937. Writing in the *Annals of the American Academy of Political and Social Science* in 1935, the president of the University of Wisconsin declared that radio (and the talking picture and, in the future, television) would transform society as much as the printing press had five hundred years earlier.[39] He hailed the technical "genius" of those who developed radio, but expressed doubts about those who programmed it. He criticized educators who were "reluctant to change" and who regarded radio as inherently shoddy. Frank urged the use of broadcast technology in formal education, but he saw even greater potential for radio to shape democracy. He predicted that the technology itself would produce "a new kind of statesman and a new kind of voter." Whether commercial or noncommercial, "the microphone is the deadly enemy of the demagogue—a ruthless revealer of 'hokum.'" He believed individual citizens sitting by their radios were more critical, and more rational, than groups of citizens hearing the same information in a public meeting. On the radio, he wrote, ideas must stand on their own without "the crutch of emotional crowd reaction." Irrational arguments and pleas to emotion, which he termed "demagogic tricks," would cause Americans sitting in quiet rooms to "laugh derisively."[40]

Frank's article quoted Gandhi: "You think your souls are saved because you can invent radio, but of what elevation to man is a method of broadcasting if you have only drivel to send out?" The UW president called on radio to soar above drivel because broadcast content educates and informs whether or not programmers intend it. His article laid out goals for educational radio that remain relevant in the twenty-first century. Radio programming, he said, needed to promote "intelligence and moral responsibility." It should seek not only a more intelligent nation but a more integrated nation, one that encourages understanding among diverse groups. He grasped that educational radio must seek out listeners who would not automatically flock to "quality." "Quality must learn to sing," he wrote, and concluded that education could "get away" with dullness if its target was a prisoner in a classroom, but not when that target could turn from dull quality to interesting frivolity with a simple twist of the dial.[41] (He expressed no opinion on the role of fiddle music in making education sing.)

In the view of the president of the University of Wisconsin in 1935, the weak signal that failed to impress those gathered in Professor Terry's living room in 1917 had acquired the power to become central to the future of democracy, to a more educated nation, and to a better understanding among diverse people. For the Wisconsin Idea to have any meaning at all, it had to use radio.

2

THE STATE STATIONS

1929–1945

O n the first of May in 1929, Professor Terry died of a heart attack. He was only forty-nine.[1] He had set WHA on a path toward public service broadcasting, but others would take it to its destination. With Terry gone and Lighty banished, President Frank split the management of WHA between Professor Edward Bennett of the Engineering School and Professor Ewbank of the speech department. Bennett would handle technical and business operations, and Ewbank would oversee programming matters. Ewbank, in turn, hired a series of his graduate students as part-time program directors. The third young man he hired proved so dedicated and effective that the Radio Committee gave him a second one-year appointment as program director and made the appointment full-time. In 1931, they promoted him to full-time manager of WHA. Harold B. "Mac" McCarty became the first person able to devote all his considerable energies to that task. The journey Terry began, McCarty continued over the next thirty-five years.

After graduating from the University of Illinois, McCarty came to the University of Wisconsin to study and teach acting. He remained a performer throughout his career. Had he not come to the university, he might have become a circus barker or ringmaster, cheerfully and enthusiastically stirring a crowd's excitement for the wondrous acts he was about to present. He was at his best presiding over meetings of Madison's downtown Rotary Club, leading songs and the Pledge of Allegiance. The Rotarian displayed a talent for diplomacy. He deftly manipulated those who disagreed with his decisions and those whose requests he could not satisfy.[2] In his first year as manager, he expressed regret to a listener in Manawa, Wisconsin, that the station could not play "When My Hair Has Turned to Silver" in honor of his mother's birthday, but expressed the hope that "your mother had a pleasant birthday and many more and extend to you and to her our sincerest regards."[3] McCarty had a gift for remembering people. Each spring at the annual "WHA Family Reunion," an event that brought the station's friends and supporters to Madison for a dinner and entertainment, McCarty would tour the room, introducing each attendee with a personal story or remembrance.[4] Some found his perpetual smile annoying, but his charisma attracted more people than it repelled, especially in his younger years. McCarty married late and had no children. He devoted his life to WHA.

Soon after he became manager, McCarty convinced university administrators that he needed a full-time assistant, and he did it in the midst of the Great Depression. He hired Harold Engel, who, like McCarty, stayed with WHA for thirty-five years. The name Engel became linked with that of McCarty throughout their tenure and beyond. Engel was a graduate student in the economics department with professional experience in the advertising industry. His strengths complemented McCarty's. He worked methodically, quietly, and effectively to build the organization McCarty publicly trumpeted. If circus barker were the alternative career for McCarty, trustworthy small-town pharmacist would have fit Engel. With the backing of Professors Ewbank and Bennett, the two Harolds tackled technical improvements and program innovations.

On the technical side, they sought a signal that covered the entire state. Ideally, they wanted a clear channel station in Madison, one that would allow them, and only them, to operate on a high-power single frequency. They had good reason to believe they should get one. The Federal Radio Commission had established a policy for fairness among the states in allocating clear channel stations to operate with sufficient power to cover vast regions and ensure service to sparsely populated rural areas. Under the commission's formula, Illinois was entitled to two, possibly three, clear channels to serve the state from Chicago. In fact, Illinois had five, at least two more than the formula allotted. Surrounding states, including Wisconsin, were each entitled to one clear channel under the formula, and

H. B. McCarty in his Radio Hall office in the mid-1930s. IMAGE COURTESY OF THE
UW-MADISON ARCHIVES, #S14733

all but Wisconsin had one. McCarty and Engel thought the commission
should reassign one of Chicago's clear channels to Wisconsin. WIBA and
WTMJ, commercial stations in Madison and Milwaukee, agreed and, along
with WHA, made one or more unsuccessful applications to poach a Chicago station for Wisconsin.[5]

While a clear channel was ideal, the more realistic approach to statewide coverage for WHA involved a station built by the Wisconsin department of agriculture and markets near Stevens Point. The agriculture
department provided the market information for WHA's noontime farm
broadcast. It decided to build its own station for that purpose in northern
Wisconsin beyond WHA's coverage area. The department of agriculture
station took the call letters WLBL and located its studio at the state normal
college now known as the University of Wisconsin–Stevens Point. Consequently, the state of Wisconsin owned two radio stations operated by

different state entities, the university in Madison and the state agriculture department in central Wisconsin. Both emphasized service to farmers and their families. Both operated on a shoestring budget. Both provided a reliable signal to only part of the state. Both operated for only a few hours a day. Would it not make sense to join forces and build *one* higher-powered station to serve the entire state?

With technical leadership from Professor Bennett and political leadership from Professor Ewbank, the university got behind the single-station idea. The state's agriculture department agreed to let the University of Wisconsin program a single new station, which would carry the market and agricultural information listeners valued. The Republican governor, Walter Koehler, got behind the idea, too, and personally led the negotiations. When Progressive Phillip La Follette replaced Koehler as governor in 1931, he threw himself into the cause with at least as much fervor as his more conservative predecessor. Thus, the state of Wisconsin sought from the federal government a license for a single high-power radio transmitter near Stevens Point. It would operate at 900 on the radio dial with the call letters WIS. In a tradeoff to keep other nearby stations off their frequency during daytime hours, the state and the university agreed to forgo postsunset broadcasting, which eliminated live coverage of evening sporting events and concerts but guaranteed reliable service to schools, farmers, and their families during the day. The state legislature provided special funding to construct the new station. State agencies interested in using the station would contribute to operating costs. The state would provide funding in subsequent years as a line item in the university or agriculture department budget.[6]

In January 1932, Governor La Follette and university president Frank led a small delegation of state and university officials to Washington to make their case to a Federal Radio Commission examiner. Frank's prepared testimony listed the reasons the commission should approve the request. The state and the university needed the radio station, he argued, to provide information on agriculture and home economics to dispersed populations, to provide continuing education to adults, to provide programming for use in rural schools, and to enhance the democratic process by providing a statewide forum for candidates and discussion of issues. Near the end of his testimony, President Frank made a more fundamental argument. He proclaimed that not-for-profit organizations committed to serving the public interest, including universities and state governments, should have broadcast facilities equal to those "placed at the disposal of private interests and private enterprise."[7] The Federal Radio Commission denied Wisconsin's request, seemingly because it rejected Frank's fundamental argument that public-interest entities like a state or university are entitled to facilities as good as those granted to commercial stations operated for profit. The hearing examiner replied that a state has no "right" to a radio facility and should compete with all applicants on the single criterion of which would offer the most "service" for the most people. If service was defined primarily as the number of people who listen, commercial entities were bound to win.[8] The commission said education and public service should receive no special consideration. This declaration forecast Congress's decision three years later not to reserve any frequencies for not-for-profit operations. The United States stood opposite the rest of the Western world. Other countries gave not-for-profit "educational" broadcasters the priority, and in countries such as Britain, public service organizations even enjoyed a monopoly for thirty years before commercial broadcasting was authorized.[9]

This preference for commercial broadcasting over not-for-profit broadcasting went largely unchallenged in the United States. Few at the time would have argued with powerful "radio priest" Father Charles Coughlin when he said, "'Like everybody else who is interested in the welfare of radio and of broadcasting, I prefer that the broadcasting stations should remain in the hands of private individuals."[10] In a blunt letter to the radio priest, the normally mild-mannered Harold Engel took exception to using "everybody" in that statement. He pointed to educators, clergy, and editors who strongly disagreed. So, he said, would the "sane minded citizens" who approved of Wisconsin's state radio "with the absence of advertising, blatant jazz and money-making schemes." Pointing to the "fabulous" sums of money commercial radio stations received from advertisers, Engel wrote, "Radio is influencing people, but who are the teachers?" He concluded that no commercial station could profitably provide the kind of programming WHA presented every day.[11]

Having failed in Washington, the state and university dropped back to plan B. WHA and WLBL would remain separate, operate on two different frequencies with less power than either needed, and share programming created mostly at WHA in Madison. Neither station would have night-time hours, but each could operate full-time during daylight hours. This arrangement depended on getting programming from WHA in Madison to WLBL near Stevens Point. They could either lease an expensive broadcast circuit from the telephone company or rely on an inexpensive, but undependable, off-the-air pickup. The two stations used each method at various times, depending primarily on their budgetary situation. Ultimately, the broadcast circuit became permanent. Rather than a "State Station," Wisconsin would operate two "State Stations" with substantial shared programming. The two stations gave Wisconsin a radio service that covered much of the state's geography. It did miss the far northwest corner of the state, however, and marginally covered the populous counties along Lake Michigan from the Door Peninsula to Kenosha. The arrangement provided a good start toward a genuine statewide service, a goal McCarty and Engel continued to pursue.

Their more immediate technical challenge, however, came from commercial stations WTMJ in Milwaukee and WIBA in Madison. Each station used the "Wisconsin deserves a clear channel" rationale to justify their expansion at the expense of the State Stations. WTMJ made its move in 1933. It made the case that Wisconsin's clear channel station should broadcast from the state's largest population center. WTMJ argued that it carried the programming the people of the state wanted, programming WHA and WLBL could never afford to produce or acquire on their very limited budget. The *Milwaukee Journal* station proposed to keep its Milwaukee transmitter and build a second transmitter on the same frequency (900 AM) in the same central Wisconsin location (Stevens Point) that the government had denied to the state and university two years earlier. If approved, WTMJ's new station in Stevens Point would force WLBL off the air and require WHA to drastically reduce its coverage. In exchange, the Journal Company offered to give the state or university two hours per week for educational programming receivable in every corner of the state.[12] The university's Radio Committee was not impressed. Professor Bennett called the proposal "ridiculous," and Professor Lighty declared forthrightly from the sidelines: "Good educational broadcasting cannot come from private competitive businesses."[13]

The proposal from WIBA was even less generous to the State Stations. *Capital Times* editor William Evjue, the self-proclaimed crusader for the public interest against private profit, made an exception for his newspaper's for-profit radio station. Evjue proposed to eliminate the entire university/state operation on the grounds that an upgraded WIBA could do what the State Stations did and much more. "Unique programs will originate from the seat of government of Wisconsin and the seat of the university, such as educational, governmental, and agricultural information and instruction; athletic and other university and state events. The service to the state and the university will be far superior to similar service now rendered by any existing station."[14] His term as governor now ended, Phillip La Follette represented WIBA against the State Stations he had championed two years before.

Defense of WHA and WLBL fell to the new Democratic governor, Albert Schmedeman, and, of course, to university president Glenn Frank. The governor argued that only noncommercial broadcasters could competently provide the services Evjue proposed. At minimum, he said, not-for-profit stations must supplement commercial broadcasting. "It would be unfortunate," the governor said, "if all advice on the use of electricity on the farm should come in the form of indirect advertising for electrical manufacturing and supply companies."[15] While the Federal Radio Commission remained unsympathetic to claims of not-for-profit broadcasters, it declined to choose between the two commercial applications and denied both. The State Stations survived.

At the same time, McCarty and Engle faced a technical challenge closer to home. The chairman of the physics department wrote to Professor Ewbank on January 14, 1932. He complained that WHA's primitive facilities in Sterling Hall caused "interference" to the department's research and teaching. Calling the impact on his colleagues' work an "injustice," the chairman concluded with a polite request that WHA "give up the afternoon broadcasting after 1:30 PM except on Friday afternoon."[16] McCarty responded by taking President Frank a list of programs that would go away if WHA complied with the physics department request. More important than losing specific programs, McCarty said, the Federal Radio Commission would give those hours to another broadcaster if WHA did not use them.

The original heating plant for the University was transformed into Radio Hall in 1934.
IMAGE COURTESY OF THE UW-MADISON ARCHIVES, #S02765

Clearly a new home for WHA outside the physics building would solve problems for both the physics department and the radio station. WHA needed much more room and a permanent home. The makeshift arrangements in Sterling Hall did not work for WHA's current activities and would never accommodate the grander dreams of McCarty, Engel, the Radio Committee, and the university's president. Tucked into the hillside behind Science Hall was the original campus heating plant. It provided the solution. Once a new heating plant replaced it, the industrial-style stone building saw a variety of uses, but none as important as providing a home for radio. The university used federal money that the Works Progress Administration provided. With Engel's guidance, the heating plant acquired a large production studio and two smaller on-the-air studios plus a rabbit warren of offices tucked into leftover nooks and crannies. The staff took over the building in 1934. A grand lobby reception room dominated the interior. Engel told visitors the Native American motif he used to decorate the lobby symbolized the oldest form of human communication in Wisconsin, while the building's state-of-the-art radio facilities represented the newest. The lobby provided the perfect location to welcome the hundreds of schoolchildren and others who visited Radio Hall each year. In 1940, a local artist painted a mural at the north end of the lobby. The mural memorialized Professor Terry and his students, Professors Lighty, Bennett, Ewbank, and Hopkins, as well as Pop Gordon and other early WHA musicians. McCarty and Engel did not include themselves in the mural.

As they made progress on their technical and facilities challenges, Ewbank, McCarty, and Engel set out to make good on the promises President Frank made when he testified, unsuccessfully, in favor of a single high-power station before the Federal Radio Commission commissioner. They built a program schedule that provided practical information on agriculture and home economics, continuing education for adults, programs for use in rural schools, and services to enhance democracy.[17]

Clearly, service to farm families took priority. 9XM's first transmission had consisted of weather and current prices paid on agricultural markets in Chicago and other Midwest locations. The state agriculture department built WLBL to provide similar information to the northern part of the state. These basics were in place before McCarty and Engel arrived. Having its own faculty advisory committee, Professor Hopkins's agricultural journalism department operated without much guidance from station management. Staff and students from "Ag Journalism" prepared "A Half Hour of Useful Information" for broadcast at noon, six days a week.[18] McCarty himself read this information as his first assignment as a part-time announcer and continued to host as program director. Later, the department appointed Professor Milton Bliss to produce and host the *Farm Show*. Weather and market information remained central to the program that also featured interviews with faculty and other experts.

While the farm program tried to sound friendly and helpful, no one embodied those qualities more perfectly than Aline Hazard. She began hosting the *Homemakers' Program* in 1933 and continued for the next thirty-two years. Seeking to promote a "more beautiful and worthwhile Wisconsin home life," Hazard greeted her listeners six mornings a week

A mural by John Stella, still in Radio Hall today, celebrates the early years of radio in Wisconsin. On the left are students and technicians. In the center are program director William Lighty at the podium and chief operator Malcolm Hanson and station manager Earle Terry seated at the table. Standing behind them, from left to right, are Radio Committee members Andy Hopkins, Edward Bennett, and Henry "Heiny" Ewbank. On the right are early broadcasters, from left to right, Waldemar Geltch, Pop Gordon, and Paul Sanders. JAMES GILL/WISCONSIN PUBLIC RADIO

with a cheerful "Good morning, homemakers!" She was always carefully groomed and dressed as if she were about to go out to lunch with the ladies. She told her audience how to clean their houses and how to set an elegant formal table. She gave advice on parenting ("Should we let our children skip grades?") and on homemaking ("Dressing up our common dishes").[19] She was "a cross between today's Martha Stewart and Oprah Winfrey."[20] Hazard not only dispensed advice; she also served as a trusted friend who helped to form thousands of isolated Wisconsin farm women into a community.[21] Quite literally, her voice on the radio was closer and more intimate than the nearest neighbor. Whether a woman lived on a farm, in a town, or in the city, Hazard was a trusted confidant. Under Mrs. Hazard, which is how everyone addressed her, the program offered more than homemaking information. Hazard regarded rural women as multifaceted

people interested in self-development and community service. She started her daily program with a music appreciation segment, read and discussed literature, and spoke to issues of concern to women beyond their roles as mothers and housewives. As her following grew, she received more than ten thousand letters a year. Wrote one fan, "Thank you so much, Mrs. Hazard, you can say everything so nice and warm-heartedly, [it's] no wonder people respond to you right away. I think you must be one of the most beloved personalities on the radio."[22]

In 1933, the same year Aline Hazard launched her homemakers' program and in the depth of the Great Depression, Harold Engel noted that approximately 120,000 young people between the ages of fourteen and twenty in Wisconsin were unemployed and out of school. In the spirit of Charles McCarthy's admonition to "make every man more efficient so that the door of opportunity may always be open before him," Engel proposed to prepare these young people for work by offering courses on the radio. The ten thirty-week *College of the Air* "courses" did not carry university credit, but did award certificates to students who signed up to receive a free study guide and passed a final exam. Each of the half-hour weekly courses was decidedly practical.[23]

In its first year, Engel's *College of the Air* curriculum covered agriculture ("Farm Life and Living") and home economics ("You and Your Home"). It included humanities courses such as "Enjoying Your Leisure," described as arts appreciation "in a popular style," science courses ("The World about You"), and social science programs ("Social Problems Today"). Other courses included instruction in letter writing, public speaking, Spanish, aeronautics, and typing. The typing teacher explained that she could give the same verbal commands on the radio that she gave in the classroom. The clickety-clack of such a radio program would not interest a casual listener not enrolled in the typing course, but it provided unique opportunities for students to acquire skills that might serve them and the broader state economy during those difficult Depression years. In its second year, *College of the Air* enrolled nearly fifteen thousand students across the state.[24]

While Engel's *College of the Air* provided employment-oriented instruction, Professor Lighty's commitment to informal liberal education continued. In the summer of 1931, WHA experimented with a technique to make radio appearances less challenging for the faculty. Rather than ask faculty to prepare special talks for radio, the station decided to take the microphone to the faculty member in his or her natural habitat, the lecture hall. It started with a summer session lecture series on music appreciation. The station asked for listener reaction because "Professor Morphy is anxious to know whether or not the radio broadcasts of the music appreciation lectures are being heard and appreciated,"[25] but the course did not offer any formal mechanism for feedback or evaluation. By the late 1930s, *College of the Air* adopted Lighty's approach and dropped formal job-oriented instruction to offer college lectures for informal listening. For the next forty years, *College of the Air* would allow listeners to eavesdrop on classrooms on the UW–Madison campus and make Professors Petrovich, Agard, and Risjord, among others, household names in homes across the state.

Relocation to the university's former heating plant provided the unexpected benefit of direct access to the system of steam tunnels that connected the old heating plant to all campus buildings. Station engineers strung broadcast lines throughout the tunnel system, turning classrooms across campus into radio studios for *College of the Air*. Listeners who heard an entire lecture series and read the books on the reading list enjoyed an

Radio was an improbable choice for teaching typing, but the original *College of the Air* gave it a try. IMAGE COURTESY OF THE UW-MADISON ARCHIVES, #S15275

excellent educational experience, but not college credit, since listeners did not enroll, did not pay tuition, and did not take exams. While few listeners actually followed courses that intently, many were taken to the college classroom, perhaps for the first time.

College courses fall squarely in the tradition of the university. But the title of "State Station" implied something more than a university station, and to help define that difference McCarty looked to the world's ultimate "state" station, the BBC in Britain.[26] Among other services, the BBC broadcast programs to children in schools, and McCarty decided to borrow that idea. Radio could provide a vital service in a state that still taught many kids in one-room schoolhouses and others in schools too small for specialist teachers, such as in art, music, and science. More a performer than an academic, McCarty would have rather spent time engaging children in the classroom than talking to adults in the lecture hall. For McCarty, as for much of the public, "education" meant kids in classrooms. He opened the *Wisconsin School of the Air* in October 1931, the year he became the State Stations' manager.

Wisconsin quickly became known as the national leader in the production and use of educational radio in the classroom. The UW broadcasters produced far more programs and reached many more students than any other noncommercial station. Its programs proved more entertaining than those produced elsewhere, reflecting McCarty's philosophy that education needed a strong entertainment element to build and hold an audience as well as convey content. Besides, programs that were fun to hear were also fun to produce. Ohio State University, in cooperation with Cleveland Public Schools, was the other leader in school programming, known for its emphasis on planning and evaluation, but Ohio had fewer programs, reached fewer students, and offered programs that were far less fun than those produced in Madison.

McCarty recruited the director of health education for the Madison public schools as his first radio teacher. He had heard her on commercial station WIBA. "The lilt in her voice and her obvious affection for children" impressed him. Fannie Steve's *Rhythm and Games* for kids in kindergarten through third grade stayed on the air for thirty-five years. When Steve retired in 1966, McCarty said she still retained "her bounce and her chuckle, her optimism and her enthusiasm."[27]

For music education, McCarty turned to the same Professor Edgar "Pop" Gordon who had provided music appreciation programs for adults in 9XM's earliest days under Professor Lighty. Aimed at fourth through eighth graders, Gordon's *Journeys in Musicland* drew the largest enrollment of all the *School of the Air* offerings and brought thousands of kids to Madison each spring for a giant sing-along in the UW Stock Pavilion.

Wakelin McNeel, known on the air as "Ranger Mac," provided the third major star in the *School of the Air* constellation. "Chief" of the Junior Forest Rangers in the Cooperative Extension's 4H program, Ranger Mac introduced students to nature studies in a world he described as "overcivilized." He began each show with an enthusiastic, "Hello, girls and boys: This is your day—so, up and away" and closed with the benediction, "May the Great Spirit put sunshine in your heart today and forevermore. Heap much."[28]

In the fall of 1932, one or more of these programs was heard in classrooms across the state, reaching twenty-three thousand students a week. By 1945, *School of the Air* offerings were heard in one-third of Wisconsin's elementary schools.[29] Pop Gordon's *Journeys in Musicland* drew the most

Fannie Steve led the weekly show *Rhythms and Games* when *School of the Air* began in 1931 and continued in this role for another 30 years.
IMAGE COURTESY OF THE UW-MADISON ARCHIVES, #S14738

students, thirty-nine thousand each week. Close behind were Ranger Mac and Fannie Steve.

School of the Air enriched education in schools around the state. It also provided the most indisputable rationale for the state to continue funding the State Stations. As McCarty observed at Fannie Steve's retirement, "Actually, there were times when the *School of the Air*—personified by those three hearty pioneers, Fannie Steve, Pop Gordon, and Ranger Mac, with their widespread popularity and approval—saved the broadcasting service from extermination."[30]

Perhaps the most surprising use of radio was the *Let's Draw* series, which taught visual art without visuals. James Schwalbach, a teacher at Milwaukee's Washington High School, tapped into students' imaginations to inspire them to create their own art rather than copying what the teacher showed them. In the spring of 1939, for example, he led a unit on capturing "feelings" in pictures, the feeling of "coldness," for example,

Pop Gordon leads the annual sing-along at the UW Stock Pavillion. IMAGE COURTESY OF
THE UW-MADISON ARCHIVES, #S14907

and the feeling of "love," the feeling of "gayety," and the feeling of "night
and darkness," abstract concepts more challenging to convey in pictures
than a house or a tree.[31]

Without question, Wisconsin's programs were imaginative, engaging,
and popular, but did they actually *teach* anything? This question had national importance, because the only federal agency that took an interest
in—and might help fund—educational radio was the US Office of Education. This federal agency was in charge of improving K-12 education.
On the state level, the answer to the question might convince Wisconsin
to provide significant funding. But for both the feds and the state, that
answer would have to confirm that educational radio was indeed educational by the normally accepted standard of K-12 education. To answer
the question, as well as to satisfy Professor Ewbank's growing interest in
using social science methods to evaluate radio programming, McCarty, in
1936, applied for—and received—substantial funding from the Rockefeller
Foundation for a "Research Project in School Broadcasting."[32] He sold it to

the foundation as a national model for such research and as a way to prove radio's educational benefits.

The study design reflected the highest standards of social science research at the time. It required each program series to define its educational objectives and specific measurable outcomes. The outcomes went beyond mere facts to include changed attitudes toward the topic or an increased interest in it. Writers worked with content experts to translate the objectives into scripts, which then went to a "production man" who organized actors, sound, and music into a finished program. The study design established control groups who would learn the same material in the radio programs from their classroom teacher. The outcomes for these groups were compared to those of the students who heard the radio programs. Finally, evaluators administered tests and collected reactions to try to understand what worked, what did not, and why.[33]

When the Rockefeller Foundation received the results of the two-year Wisconsin project, it declared the results "mixed."[34] While results were positive overall, none were "statistically significant," that is, the result was as likely to have happened by chance as by the teaching method. More disappointing, the control groups taught the material by traditional teachers tested slightly better than those who heard the radio program, although not sufficiently better to have "statistical significance." Yes, students got something out of the radio programs, but no more than they would have from a traditional classroom teacher. Because of such results, the Rockefeller Foundation withdrew funding for broadcast educational programming.[35] State Station advocates chose to highlight the good news that came from the research effort. Students *did* learn from radio (even if traditional methods worked just as well). More important, teachers liked the programs as "supplements" and "enrichment" to their teaching. In the years that followed, it was that message McCarty and Engel took to the university administration and the state legislature in their campaign for greater support.

While the State Stations called all of their programming "educational," some of their most important innovations had little to do with college or K-12 classrooms. Instead, they emerged from the broader vision of the Wisconsin Idea as McCarthy, Van Hise, and La Follette had described it. Educational radio contributed to the quality of democracy in Wisconsin.

President Frank's statements of purpose held out the hope that the State Stations "may enable Wisconsin to re-create in this machine age the sort of unhampered and intimate and sustained discussion of public issues that marked the New England town meeting and the Lincoln-Douglas debates."[36] After meeting with the university's Radio Committee, Professor Ewbank proposed the stations produce radio forums to bring together proponents of different viewpoints on controversial topics. The professor displayed his national reputation as an authority on "rhetoric" and argument by distinguishing between "facts" and "the inferences to be drawn from the facts," which he saw as the point of such debates. The State Stations, he said, should "not assume responsibility for the accuracy of anything said in these forums." The stations' obligation ended with ensuring representation from all sides of the issue. Such debates should take up all types of controversial issues of public importance whenever they arose in the state.[37]

As chief advocate of the "discussion and debate" function of Wisconsin's State Stations, Professor Ewbank elaborated his views in a five-page letter to Governor La Follette in 1934. It followed up on "our discussions" (presumably between Ewbank and the governor) about using radio in political campaigns. Ewbank proposed to offer equal time to all parties with candidates on the statewide ballot. He explained to the governor, "This will equalize opportunities among rich and poor parties."[38] Once again, the State Stations looked to the BBC for its model. In Britain they called it *Party Time*; in Wisconsin, McCarty and Engel called it *The Political Education Forum*. As in Britain, they would give party officials complete control of content, including decisions on participants. The stations would have no say in the content and expressed their faith that "candidates will avoid language problems."[39]

While Ewbank wrote to the governor in 1934, the State Stations had offered time to political campaigns in 1932, an event reported—and praised—by the *New York Times*.[40] In the summer of that year, representatives of the Democratic, Prohibition, Republican, Progressive Republican, and Socialist parties joined Engel for lunch at Madison's Lorraine Hotel to sign an agreement he had drafted for the forums that fall.[41] The agreement began with the hope that "if each party is allowed equal opportunity to present its case over the State Stations, the voter can get a much more adequate understanding of the issues and can cast a much more intelligent ballot."[42] WHA and WLBL each set aside thirty minutes at the noon hour and thirty minutes in the late afternoon for candidates to speak to voters. Even the state's tiny Communist Party had its turn, the speaker concluding his half-hour talk with, "Organize the mass struggles of all toilers against the capitalist program of misery and hunger."[43] Engel's document dismissed the charge that giving free, uncensored time to all parties might be dangerous. "The process of avoiding danger often results in avoiding progress of any sort," he wrote.[44]

The *Political Education Forum* continued for every statewide election for the next thirty-six years. Complementing the free and uncensored time for candidates was a similar opportunity for all members of the legislature "to speak to the folks back home." Engel's 1941 invitation to appear on the *Wisconsin Legislative Forum* assured legislators that those folks back home "want to hear from their representatives in Madison."[45] Those who accepted Engel's invitation received fifteen minutes of free broadcast time to send the message of their choice. They also received training sessions from Engel on "the ABCs of broadcasting." The first training session centered on radio writing, "timing, tricks in writing, analyzing the audience, holding the listeners, persuasiveness, intimate style, vocabulary, sentence length and structure, slang, illustrations, and a strong close." He followed the writing session with advice on radio speaking, "relieving mike fright," voice analysis, and suggestions for improvement.[46]

The state provided studio space in the Capitol for the convenience of legislators and the State Stations. The Capitol facility fell short of the normal standards for a studio, but no other media outlet had facilities in the Capitol. A door off a staircase landing near the Assembly Chamber gave access to a claustrophobic, low-ceilinged room that might otherwise have been of little use other than storage. It did have a small window, however, plus soundproofing, a table, a microphone, some chairs, and a broadcast line back to Radio Hall on campus. Whatever its practical shortcomings, the Capitol studio sent an unmistakable message about the central role of the State Stations in government service to the people.

A genial Harold Engel greeted legislators at the time of their broadcasts, made them comfortable in the studio, provided advice and encouragement to the neophytes, told them when to start and stop, and subtly reminded them of the importance of the State Stations to the political system and their individual careers. His genuinely down-to-earth style allowed Engel to exchange fish stories and chat about other topics that lubricated a comfortable rapport with the legislators. An invitation to visit him at his cottage on Lake Wisconsin was likely to conclude the encounters. Engel might also mention the place of the stations in the state budget and coverage gaps that new or improved facilities might address.[47]

This unique service became an issue in 1938 when WHA boldly applied to steal the clear channel of Chicago's WMAQ. Not surprisingly, the *Chicago Tribune* editorialized against stripping the city of one of its most important broadcasters. More surprisingly, however, the *Tribune* took the occasion to thunder against the Wisconsin State Stations as "dedicated exclusively to the advancement of the political fortunes of the La Follette machine."[48] The editorial conceded that Wisconsin's *Political Education Forum* gave no more time to the La Follettes than to their political rivals, but claimed "consistently, in campaign season and out, the station is used to promulgate the social and political ideas of the La Follettes and to extol legislation that they support."[49] The editorial provided no examples of such bias. McCarty and Engel were studiously nonpartisan in their own speech and behavior and stuck to the "all sides of the issue" approach to political and governmental matters. However, the issues on which they provided all sides may have been those raised by La Follette partisans, who did, after all, drive the political agenda of the state. More fundamentally, the very existence of such a state-operated forum conformed to the Progressive agenda of government reform and stood in marked contrast to the opinionated approach of Colonel Robert McCormick and his *Chicago Tribune*.

President Frank did not mention "entertainment" in his initial testimony on behalf of educational radio before the Federal Radio Commission examiner in 1931, but his subsequent testimony and speeches promised to provide music and drama to serve the "recreational needs" of the public. Broadcasts of athletic events were more entertaining than educational, of course. The same could be said for the high school and college bands and choirs, variety shows, and other entertainment that the State Stations aired, or of the fifteen minutes of marching band music that launched each broadcast day.

Engel assured commercial broadcasters that the State Stations did not duplicate programming on other stations, but Professor Ewbank observed that the line between entertainment and education was blurred. What is entertainment to some is education for others. Its monopoly status allowed Britain's BBC to engage in "cultural uplift," forcing Britons to listen to material "they did not know they wanted," that is, a brow level higher than they would have preferred. BBC listeners were "educated" in music, drama, and other entertainment that was "better" than what they might have chosen had they been given a choice. Without the BBC's monopoly, WHA faced long odds in introducing audiences to material they did not know they wanted, leaving it with the less exalted role of primarily serving audiences that already wanted classical music and other "high culture" material.[50]

Although rightly remembered for his leadership in "educational" programming, McCarty never lost touch with his origins as an actor and entertainer, and he attracted to his staff individuals with a similar sense of performance. *Chapter a Day,* the longest-running program on the State Stations and one that continues today, epitomized the blending of education and entertainment. The books McCarty and his successors read on the air might educate, but first they had to entertain enough to attract and hold listeners. As with programs for children, McCarty believed, programs for adults should educate as they entertained. Some programs, of course, had little educational value but attracted listeners who might never listen to the stations otherwise. Play-by-play sports provide the most clear-cut example, a lure to try some of the State Stations' more substantial offerings. Sports were one of the few areas of conflict between McCarty and his normally passive partner in educational broadcasting, Wisconsin's department of agriculture. Because it wanted to use them to promote agricultural products to a broad audience, the ag department, not the university, paid the costs of sports broadcasts and felt it "owned" them. In 1937, the head of that agency accused the football announcer of rooting for the Badgers too much and for making too many mistakes. He added that McCarty's color commentary on those broadcasts was "too neutral" about agriculture. The department head suggested that McCarty remind listeners during the sports broadcasts that agricultural products make nice Christmas gifts.[51]

In less than a decade, McCarty, with Engel, had built the State Stations into a unique institution in American broadcasting and the undisputed leader in noncommercial radio programming for agricultural extension, school programming, adult education, public affairs programming, and entertainment. It was the best among university broadcasters nationwide, but nowhere near good enough, according to Charles Siepmann, a top BBC program executive. He was sent to America in 1937 to assist the Rockefeller Foundation in evaluating and fostering noncommercial broadcasting in the United States. Siepmann visited every significant university broadcaster in the country and found all of them sadly wanting, certainly compared with the world standard, his own BBC. In most places he found pathetic facilities, small, unprofessional staffs, and little support, financial or moral, from university administrators.

In Wisconsin, however, he was pleased with the facilities at Radio Hall and praised the university and the state for their unique support of educational radio. But, Siepmann reported to the Rockefeller Foundation, "McCarty is a disappointment. He speaks the right language but his values seem to me to be derivative."[52] That McCarty had derived his key ideas from the BBC itself did not seem to impress Siepmann. The BBC executive observed "a somewhat sensitive vanity" in McCarty. He concluded that McCarty's "conception of broadcasting is emotional rather than intellectual," an observation consistent with, if more bluntly stated than, the opinion of others who worked with him. The cheerleader-in-chief's unrelenting enthusiasm actually hurt the quality of the station, Siepmann said. "I heard a performance which was not bad at an amateur level, but I reckon there is some danger of this student having his head quite turned by McCarty's excessive praise and lack of critical guidance."[53] (A revision of the WHA announcer's manual in 1940 cautioned broadcasters to mute their praise when presenting performances by nonprofessional musicians.) Siepmann questioned McCarty's professionalism and his ability to control his staff. He took issue with his ability to judge people, a reference, perhaps, to McCarty's fair-haired boy, Gerald Bartell. Bartell was a young man of considerable talent as an actor, announcer, and "production man." He provided the "juice" at the station, but he disappointed his mentor by leaving WHA to build one of the country's more lucrative commercial media empires of broadcast stations, downscale magazines, and in-room motel movies.[54]

Siepmann found Engel "competent, charming and keen," but unprofessional with little understanding of education, a characteristic endemic

Gerald Bartell, center, provided artistic and production leadership to the WHA staff under McCarty, but he left to build a commercial media empire. IMAGE COURTESY OF THE UW-MADISON ARCHIVES, #S14675

to the entire staff. He criticized the station for not paying speakers, not rehearsing correctly, and conducting classroom lectures "in an unsound manner." As for McCarty, "He protests too much, and considering the potentialities open to him at this university, less has been achieved than I should have thought was possible, even within the pitiable restrictions of finance and staff that obtain here." Nonetheless, he concluded, "If regional broadcasting is to be developed in the Midwest, Wisconsin offers the best opportunity."[55]

3

THE STATE RADIO NETWORK

1939–1965

I n 1939, Governor Julius Heil summoned Harold B. McCarty and Harold Engel to his office to meet with two lawyers and a vice president from NBC. The two Harolds had attracted NBC's attention by applying for WMAQ's 50,000-watt clear channel frequency. WMAQ was NBC's Chicago station. It blanketed much of the Midwest. McCarty and Engel argued that Wisconsin deserved a clear channel station under Federal Communications Commission policies and that Chicago had more channels than those policies allowed. They were confident of their case, but the chances of wresting the clear channel from Chicago, slim at best, slipped to zero after that meeting in the governor's office. Engel recalled the NBC vice president shaking his finger in the governor's face and saying, "Governor, we don't like this and we're going to lick you. We've got three million dollars to spend on this and we're going to lick you." The governor took offense at the messenger's tone, but he got the message. The state Senate tabled the legislation that the Assembly had passed to fund the 50,000-watt station in Madison.[1]

McCarty and Engel lamented that closed door, but the FCC opened a window. The same year Wisconsin ended its quest for WMAQ's AM clear channel, the FCC designated a set of very high frequency channels for FM (frequency modulation) broadcasting. Unlike its action in the early 1930s that, in effect, turned over the standard AM (amplitude modulation) band to commercial interests, the commission reserved 20 percent of these FM channels for noncommercial educational broadcasters. Lobbying by

Wisconsin and other advocates of educational broadcasting played a role in this decision, but so did the lack of interest commercial broadcasters showed in the FM band. Commercial stations had already invested heavily in AM broadcasting. AM radios were found in virtually every household. Commercial broadcasters asked why they should invest in stations no one could receive without buying new and relatively expensive radios, stations that would have minimal coverage compared to the multistate possibilities of large AM stations. Moreover, they saw television as the future. Television, they decided, was a potential gold mine for broadcasters and manufacturers. They would invest in television, not FM radio. Therefore, they had little reason to oppose a noncommercial foothold on the FM dial.

C. M. Jansky, a former student engineer, had already sold McCarty and Engel on the virtues of FM before the FCC acted. McCarty wrote, "The State of Wisconsin looks to FM in the hope that it has found the answer to its radio prayer [to blanket the state]."[2] Recognizing that they needed support beyond the university to fulfill their ambitions for such a network, the Radio Committee successfully lobbied the university's Board of Regents to create a State Radio Council. The council would have eleven members: the governor, the secretary of agriculture, one representative from the state normal schools, one from the technical colleges, one from the department of public instruction, the university president, and five other UW administrators. Professor Ewbank added chairmanship of the State Radio Council to his duties as chair of the university's Radio Committee.[3]

The newly created State Radio Council began its work in 1939 with a philosophical statement about the potential positive uses of radio. It issued this statement at the same time Hitler was using radio to inflame the German people. Reflecting Ewbank's strongly held views, the council said it would dedicate the new agency to "democratic ideals and principles" and, following the Wisconsin Idea, the "sifting and winnowing" of all ideas. The council promised to "hand [over] the best in the established patterns of thought and to prepare the way for the orderly correction of the recognized shortcomings of the present order." The council also promised not to neglect "the cultural aspects of individual and community life." Appreciation of the arts, the council said, depends "upon an acquaintance with them, and the level of tastes and interests can be raised only as the finer things become more generally available."[4]

Then, World War II stopped everything.

Engel went to the Philippines as an army officer and found the experience so interesting that he considered remaining in the military after the war. McCarty was much less happy in the war information office in New York. He wrote to Ewbank, "Six months of uselessness and frustration is just about all I can take at this point."[5] His heart and mind never left Madison; he worried about WHA and, particularly, about the loss of key staff members to the draft. Back home, the station continued under the genial, if lethargic, direction of William Harley. Harley was the heir to enough personal wealth that he could consider educational radio a pleasant avocation rather than an all-consuming career as McCarty and Engel did.[6]

All broadcasters, including the State Stations, devoted significant airtime to the war. The war gave the stations the opportunity to report current news for the first time.[7] WHA subscribed to a wire service to bring its listeners up-to-date war bulletins. At $5,400 a year, the wire service and staff to use it stretched the stations' budget and would lead to criticism about duplicating services already available from commercial broadcasters, but the Radio Committee argued that WHA had to cover news of the war. The State Stations also produced technical training programs for the navy and the army air force. They framed programs on nutrition and fitness as part of the war effort and produced programs like *Meet Our American Allies* and *Wisconsin Alert*.[8]

While the State Stations emphasized war programming, Ewbank, leading the Radio Committee, refused to let the stations become a vehicle for propaganda. In a special statement titled "Controversial Issues During Wartime," the committee acknowledged that the war atmosphere would lead to criticism "for permitting the presentation of minority points of view," but reiterated its commitment to "free and critical inquiry into controversial problems of general concern to the citizens of Wisconsin."[9]

McCarty returned to Madison before the end of the war and quickly received authorization from the regents to plan a statewide FM network. The regents had established the State Radio Council in 1938, but now McCarty decided that it needed more independence from the university. Such independence, he thought, would help gain the support of legislators suspicious of the elitists on the Madison campus. As an independent state agency, the State Radio Council would have its own budget directly from the legislature, a staff, and the authority to build the FM network. All of this was in addition to what the university already spent from its own budget on WHA. To make this bifurcated structure work, however, the State Radio Council needed to appoint the university's broadcasting director—McCarty—as its executive director. Legislation to separate the State Radio Council from the university and to build a state FM network reached the legislature in time for its spring 1945 session, just as the war was about to end.[10]

Everything seemed headed for easy approval when a half-hour student-produced program struck fear in McCarty's heart. A *UW Roundtable Student Forum* aired in March. The program discussed birth control and generated six phone calls in protest, three from individuals who had not actually heard the program. McCarty explained to the university president that a student committee had planned the program, which involved four students, two of them Roman Catholic. He said all four participants had agreed to keep religion out of the discussion.[11] McCarty defended the discussion in a lengthy letter to the Reverend William Mahoney. He described it as free of "bias, unfairness, or misrepresentation." Nonetheless, he expressed "genuine sorrow" that the program had been broadcast, calling it a mistake in judgment, a reflection of "human frailty."[12]

The university Radio Committee set up a subcommittee to study other complaints from Catholics about a freelance book reviewer and the popular *College of the Air* lecturer Professor Walter Agard. While some on the committee thought the word choices of the reviewer and Agard were "immoderate," the members supported McCarty's desire to refrain from censoring the comments of speakers, particularly the comments of a professor during a classroom lecture. McCarty did agree to talk privately with Professor Agard, however.[13]

Whether or not it was necessary to end the teapot-sized tempest, program director Walter Krulevitch turned in his letter of resignation one week after the student birth control broadcast. He took responsibility for the program and expressed the hope that his departure would end any controversy that might endanger the FM network legislation.[14] Krulevitch was a gifted classical music announcer with a great voice and an intimidating knowledge of the classical repertoire. Hence, his resignation might not have been as self-sacrificing as it seemed. Krulevitch was preparing to leave Madison for a distinguished professional and academic career in New York and Los Angeles, where his impressive "voice of God" made anonymous appearances in popular entertainment.[15]

The legislation passed in May 1945. It created and funded the independent State Radio Council to build and operate the state network, but it left the administration and programming of the stations within the university.[16] Not surprisingly, the council appointed H. B. McCarty as its part-time executive director. He assumed the directorship in addition to his university duties as head of broadcasting. In practice, the State Radio Council seldom met and did little, but with it, McCarty and Engel had created an ingenious structure, partially within and partially outside the university, that empowered them to build the only state educational radio network in the country.

The 1945 legislation provided funding to build the first two stations in the projected network, WHA-FM in Madison and WHAD on the highest peak in southeast Wisconsin near Delafield in Waukesha County. WHAD could cover Milwaukee and surrounding areas. WHA-FM carried the same programming as WHA and WLBL on the AM band during daylight hours and additional programming funded by the State Radio Council after sunset. WHAD picked up WHA-FM off the air and rebroadcast the signal in the Milwaukee region. Subsequent appropriations would add stations at the rate of one per year until the network covered the entire state in 1954.

WHA-FM took to the air on McCarty's birthday, March 30, 1947. Gerald Bartell produced a half-hour inaugural program in a highly traditional, scripted documentary style. Governor Oscar Rennebohm and UW president Edwin Fred spoke from scripts written for them to mark this unique state and university collaboration. Professor William Lighty came back to the station he helped launch and remembered the pioneering work of his late colleague Professor Terry. He then quoted the words of Charles Van Hise, father of the Wisconsin Idea, about spreading "the beneficent influences of the university to every home in the state."[17] Lighty said the new FM network would bring the state closer to Van Hise's ideals of lifelong learning and "universal enlightenment and understanding."[18]

Commercial broadcasters did not see the potential for FM when McCarty and Engel proposed the state network in 1945. Only after WHA-FM and WHAD reached the air did the commercial broadcasters

Program director Walter
Krulevitch resigned as the
legislature considered
creating the state FM.
IMAGE COURTESY OF THE
UW-Madison ARCHIVES,
#S14491

in the state recognize the potential competition. In October of 1948, the
League of Wisconsin (Commercial) Radio Stations objected to expanding the state FM network beyond those two stations. They appointed a
committee to meet with Governor Rennebohm to discuss "this wasteful,
socialistic trend." Commercial stations, they said, could carry important
educational programming without cost to taxpayers. The governor responded by removing from his budget funds to build four new stations for
the FM network, fully expecting the legislature to restore them "as they
always do." The legislature did indeed restore the four stations, and the
business-oriented governor pondered vetoing them.[19]

In response to the threat, McCarty took to the air and generated nearly
a thousand individually written letters in support of extending the network
to Green Bay and Rib Mountain near Wausau in 1950.[20] Professor Ewbank
conceded that some State Stations' programming occupied a "twilight
zone" between education and entertainment and invited the commercial

Professor Glen Koehler, left, technical advisor to WHA, and chief engineer Jack Stiehl, right, supervise the erection of an FM tower behind Radio Hall. IMAGE COURTESY OF THE UW-MADISON ARCHIVES, #S14686

broadcasters to set up a committee to discuss where they would draw the line. They never responded to his invitation. In reality, their main concern was the broadcast of UW sports. The State Radio Council's policy of allowing commercial stations to rebroadcast any of its programs, including sports, helped dissipate the opposition, and the commercial stations backed down.[21]

Ultimately, Harold Engel sealed the deal for legislative support by asking legislators in regions that did not have coverage why their constituents should not have the same services as those folks down in Madison and Milwaukee. His question generated the expected answer, and a state FM station became a tasty bit of bacon to bring home to the voters. They proceeded to cut the deals and roll the logs necessary to fund the construction and operation of the complete nine-station FM network, culminating with WHHI at Highland and WHSA in the Brule State Forest. McCarty sent a "personal note" to Rennebohm acknowledging the ambivalence the governor felt about building a state network and promising that his signature on the legislation "will redound to your every lasting credit."[22]

Opposition, however, didn't end. The 1954 Republican State Convention called for abolishing the State Radio Council and its stations because they were "socialistic," and in 1955, members of the legislature's Joint Finance Committee proposed to do just that.[23] Engel went all out to rally the forces to save what he and McCarty had built. He organized groups of listeners and met with them in their regions. Classical music listeners in the Milwaukee area comprised the largest and most committed group to come to the rescue of the FM network. An "elite" group, they had the connections and savvy to exercise influence. Letters poured in from around the state, and after public hearings in April 1955, Joint Finance allowed the proposal to die.[24] The state FM network had insinuated itself into the daily lives of many Wisconsinites, and few, even commercial broadcasters, seriously wanted to abolish it.

At its first meeting in 1945, the State Radio Council had outlined its goals in a document titled "Aims of Educational Broadcasting." Surprisingly, it mentioned neither in-school programming nor formal education for adults. Rather, it addressed Professor Ewbank's priority, "fostering democratic practices and ideals."[25] The council promised to promote free and critical inquiry into general public concerns, to present divergent views, but remain strictly nonpartisan. Recognizing its wartime entry into news reporting, the council said its newscasts would deal "only with news of political, civic, economic, or social import (not trivia nor sensation). News should seek to clarify significant news events and trends and to make more comprehensible the day-to-day news dispatches."[26]

McCarty expanded on the council's aims with a "prospectus on programming for the proposed state FM network." The document stated forthrightly what had always been his philosophy that "though produced amid academic surroundings, the programs are not permitted to become 'high brow.'"[27] McCarty's prospectus differed from the aims adopted by the State Radio Council by giving equal emphasis to education and what he called "public service." His "education" category included *School of the Air* and *College of the Air.* "Public service" covered everything else.

By the mid-1950s, *School of the Air* claimed six hundred thousand enrollments a year, although many of those students listened to more than one program and, therefore, were counted more than once.[28] From its beginning in 1931 through the early 1970s, *School of the Air* occupied a special place at the State Stations as well as in H. B. McCarty's heart. While male voices dominated on the air, women wrote the scripts and determined the content of school programming, mirroring, of course, the female-dominated elementary classrooms around the state. These women worked for McCarty quite separate from the main body of the WHA staff, but they were highly respected by those who needed their scripts to produce the programs that gained national acclaim for WHA.[29]

School of the Air did not change its fundamental purpose, approach, or programming from its origins, but inevitably key personalities moved on. Declaring in 1954 that "20 years is long enough to be on the radio," Ranger Mac passed his fifty thousand weekly classroom listeners to Professor Bob Ellison, who shed the "chief" persona in favor of a more unadorned exploration of the natural world in *The Wonderful World of Nature.* Later, retirement saw Pop Gordon's *Adventure in Musicland* and its annual sing-along in the university stock pavilion give way to the studio-centered *Let's Sing,* featuring, among others, Madison elementary school principal Norman Clayton.

While McCarty classified *College of the Air* as "education" rather than "public service," it required no registration, no examinations, and no papers, and it offered no credit. Enthusiastic letters from "housewives"—the assumed audience for these classroom lectures—said they listened for pleasure. Even without credit, *College of the Air* proved the most extensive "liberal education by radio" ever offered by WHA or any other radio station. The introduction of audiotape allowed student engineers to record lectures in classrooms not wired to Radio Hall through the university steam tunnels. More important, it allowed modest editing of the lectures,

State legislators had this image in mind when they approved the FM network in 1945.
IMAGE COURTESY OF THE UW-MADISON ARCHIVES, #S05822

eliminating some of the dead air and distractions, although plenty of in-
audible questions from students, coughs, paper rattling, passing sirens,
and ringing bells left no doubt that listeners were eavesdropping on a
classroom. With the addition of nighttime hours on the FM network, *College of the Air* settled into a three-times-a-day pattern, a morning lecture,
an afternoon lecture, and an early evening lecture.

McCarty's "public service" category encompassed what would come to
be known as "public broadcasting." His prospectus described three types of
public service programs: training for citizenship, vocational information,
and a broad "liberal education by radio."[30]

"Training for citizenship" grew most directly from the Radio Council's "Aims" document and was most uniquely "Wisconsin" compared with
other educational radio stations of the time. The phrase put an educational
spin on public affairs programming, including, of course, the legislative
and political education forums launched in the 1930s. While not essentially different from what the State Stations were already doing, this type
of programming proved the trickiest for the Wisconsin State Radio Network, as it was directly funded by the legislature through the State Radio
Council. The *University Forum,* a series that a student committee of the

Wisconsin Union produced, drew fire soon after the legislature approved building the FM network. The format gave three guests ten minutes each to speak on a controversial issue followed by fifteen minutes of open debate and fifteen minutes of questions from students. The spring 1946 season considered ten controversial issues, including "Do we need strike control legislation?," "How should we attack the housing problem?," and "Can we get along with Russia?" Such debates were exactly what the State Radio Council promised in its "Aims" document, but those who took issue with some of the views expressed—or perhaps, not expressed—complained to the university president. He defended the program but appointed a part-time faculty director and moderator for the forum.[31] Subsequently, the UW regents amended policies to assure that "wherever practicable, divergent views of controversial issues will be presented on a single program."[32] Thus the student-produced forum gave way to the *University of Wisconsin Roundtable*. News director Roy Vogelman produced and moderated the program. It did exactly what the regents' policy requested. Multiple well-credentialed experts joined Vogelman for hour-long discussions, challenging one another on the broadcast under Vogelman's neutral guidance. While some no doubt objected to the "liberal" views of UW faculty experts, few could blame the radio stations for those views or accuse them of deliberate bias.

In the spring of 1963, Governor John Reynolds presented McCarty with another journalistic challenge. Reynolds's predecessor, Gaylord Nelson, had regularly prepared a fifteen-minute governor's report for broadcast on the state radio network. Reynolds chose not to continue that practice and suggested that WHA tape and excerpt his press conferences instead. McCarty objected, stating that doing so would take too much staff time. More important, he did not want to take responsibility for the necessary editorial decisions. He was comfortable letting public figures say whatever they wanted under the stations' long-standing "forum" philosophy, but not with his staff deciding what the public should and should not hear. When Reynolds refused to reinstate the *Governor's Report* format, McCarty asked—privately, of course—"Is he stupid? Or pig-headed? Or both?"[33] Perhaps McCarty did not recognize that fifteen minutes on the radio did not mean as much to a politician in the 1960s as it had in the 1930s.

Acquiring a wire service and adding newscasts in the early 1940s moved the stations into the realm of news reporting. Staff read news provided by the United Press or the Associated Press. They did not do original reporting for which they would need to take responsibility. Their editorial judgments were restricted to deciding which wire service stories to feature. In addition to reading wire service news copy, Vogelman read commentary and analysis from print publications each morning on *Views of the News*. As with the wire service stories, none of these comments originated with WHA staff. Vogelman made editorial judgments only in deciding which opinion pieces to include, seeking to provide a range of mainstream opinions and, in the State Stations tradition, letting the listener decide.

Reading wire copy and the views of others proved tenable during the relatively quiet 1950s and early 1960s, but it became embarrassing in the mid-1960s when social movements engulfed the nation, including Madison and its university campus. Vogelman tried to ignore the clashes taking place just outside the doors of Radio Hall as he read what the *Wall Street Journal* or the *New York Times* had to say about student demonstrations at Columbia or Berkeley. When Father James Groppi and others interrupted the governor's State of the State address to the legislature, Vogelman, instead of staying with the breaking—and unpredictable—news event, abruptly ended his live broadcast from the Capitol and returned to Radio Hall.[34] That decision engendered criticism among some younger staff members, but there is no reason to believe McCarty himself disagreed with it.

By the time the FM network was established, "vocational information," McCarty's farm and homemakers programs, no longer provided the primary justification for the State Stations, but they still comprised about 12 percent of the daily schedule.[35] The agricultural journalism department affirmed the importance of these programs by assigning senior faculty member Maury White to take over the farm show from senior faculty member Milton Bliss, about the time the State Stations expanded into the Wisconsin State Radio Network. Of course, the department continued to provide the services of Aline Hazard for the *Homemakers' Program*, and department chair Andrew Hopkins resolutely supported what most listeners still called "the State Stations."

McCarty's third public service goal, "a liberal education by radio," emphasized music, drama, and the arts. In the most ambitious effort to provide such a liberal education, WHA music director Libby Monschein collaborated with her husband, Music School professor Robert Monschein, to produce *Music in Context*. The state FM network devoted Tuesday evenings in 1962 and 1963 to "a series of concerts and lectures tracing the development of western music and related arts."[36] Each evening began with two half-hour talks by university faculty. These specialists "in history, art, drama, classics, and philosophy" set the historic and cultural context for a period in music history. Those two talks led into a two-hour concert of music hosted by an expert in the music of the period under discussion. One evening began with a talk about Freud by Professor George Mosse, followed by a talk about Romantic painting by the artist John Kientz, leading into two hours of nineteenth-century chamber music hosted by Professor Robert Crane. In all, fifty different faculty members participated in the series, several of whose names live on in campus buildings, such as Helen C. White, Ronald Mitchell, and Mosse.[37] It was a liberal arts tour de force for anyone prepared to devote three consecutive hours each week for thirty-four weeks to this not-for-credit experience.

Few were that committed, of course. Far more listeners broadened their music horizons with one or more of the three classical music concerts broadcast each weekday, concluding with the two-hour FM concert each evening. The FM network transmitters provided concert hall quality to music far beyond the static-filled quality of AM radio. FM was a natural for classical music, and the state network took advantage of its strength. Programmers planned the concerts according to a formula Don Voegeli, once the station's music director, had devised. They wrote scripts under the dictum of another music director, Cliff Roberts. They were admonished not to tell more than the listener wanted to know.[38] Roberts and announcer Ken Ohst usually read the scripts on the air, although sometimes student announcers, who knew nothing about classical music, handled the task, albeit less gracefully than their professional counterparts.[39]

In all, classical music comprised 25 percent of the FM network schedule and "light" music another 15 percent.[40] Lightest of the light were the marches that launched the broadcast day with *The Bandwagon*, as they had since the 1930s. Whatever the popularity of the marches themselves, listeners loved host Bob Homme's *The Bandwagon Correspondence School* and its slightly off-the-wall faculty portrayed by chief announcer Ken Ohst. Close friends off air as well as on, Homme and Ohst ad-libbed interviews

Ken Ohst records a program with control operator Charley Farris. IMAGE COURTESY OF THE UW-MADISON ARCHIVES, #S14723

with Dr. Russell Grouse, head of the ornithology department and expert bird-watcher, and Dr. Winston Vane, chief meteorologist explaining weather phenomena.[41] Each spring, news director Vogelman joined the merriment as reporter Stark Raving broadcasting live from the last ice floe melting into Lake Mendota, with a predictably wet ending.

The Bandwagon began each morning at 7:16 with *The Weather Roundup*, a Wisconsin curiosity made possible by the unique network of FM transmitters and their operators across the state. Using its off-the-air pickup system, the duty engineer at each transmitter took turns describing weather conditions and temperature at his hilltop location. In contrast to the golden voices with a standardized cadence McCarty hired at Radio Hall, the station engineers sounded much like other people in Wisconsin, with their regionalisms and remnants of their ethnicities. Some were clearly uncomfortable with their "not in my job description" duties and mumbled through their reports as quickly as possible, while others reveled in their thirty seconds of stardom. All charmed listeners with their authenticity while providing important information to everyone who endures Wisconsin's ever-interesting climate.

Another idiosyncrasy of the state network system added to the sense of tradition emanating from Radio Hall. Engineers around the state needed to play recorded station identifications at least once each hour. Understandably, these engineers were not always sitting in their assigned seats paying strict attention throughout their shifts. If nothing else, they needed to heed the call of nature, or perhaps experience the nature that surrounded them away from the heat-generating transmitters they tended. To alert the operators to an upcoming station break, announcers in the studio struck a series of four chimes. The announcer's manual told them the "proper order for ringing chimes is 1,3,2,4. Hit them evenly and firmly, but not too hard; then quiet with the hand before making identification."[42] Somewhat later, a three-chime sequence triggered by the push of a button replaced the four-chime manual sequence, and that remained a Wisconsin State Radio Network signature long after NBC retired its similar three-chime sequence.

McCarty and Engel saw the state FM network as their crowning achievement, yet they stayed in their respective positions for another thirteen years after its completion. Staff members recall McCarty as an increasingly remote figure during the late 1950s and early 1960s, and Engel as even more remote. In July 1964, production manager Karl Schmidt wrote to a former colleague, "Mac is still on his horse here and it ain't charging anymore, even at windmills. He seems quite saddle sore and weary although as toothy as ever."[43] Except for an occasional high-minded pronouncement, McCarty seldom interacted with working staff members, who often expressed frustration with the lack of leadership while simultaneously appreciating the freedom it afforded. Although station leaders from Terry to McCarty had vowed that jazz had no place on the State Stations, McCarty passively allowed announcer Ken Ohst to produce a jazz show. Other staff carved out similarly comfortable niches. Few staff worked any harder than he or she wanted to, nor did they take on projects that did not interest them.

More than anyone else, program director Jim Collins helped fill part of the void McCarty created. A southern gentleman who chose not to live in his home region, he spent much of his life in Europe before finding his way to Madison. Collins brought with him sophistication well above that of his Badger-bred colleagues. He encouraged series such as *Music in Context*, and he brought top Madison faculty to *College of the Air*. He integrated talks, readings, and appropriate music into an hour-long magazine program of ideas called *Kaleidoscope*, and he created a more light-hearted half-hour program called *Etcetera*, built on humorous readings from *The New Yorker* and similarly urbane sources.[44]

In spite of McCarty's admonition that programs should never sound "high brow," the on-air style of the stations remained formal and "professional" compared to the sound of commercial radio in the 1950s and 1960s. That on-air sound contrasted with the more frivolous atmosphere in Radio Hall, particularly among the many part-time student employees who carried out most routine functions and were often bored by the content the stations presented. Nearly every former staff member tells stories about practical jokes they played, most of which seemed to involve trying to

Students perform on WHA. Bill Siemering, far right, went on to serve as the first program director of National Public Radio. IMAGE COURTESY OF THE UW-MADISON ARCHIVES, #S14676

make their fellows laugh inappropriately on the air. One time, for example, someone taped the receiver down on a studio telephone while fastidious program director Jim Collins hosted a live program. When the prankster called the studio number, Collins was unable to stop the ringing, to the great pleasure of the crew who thought the distracting phone was more interesting than whatever material the host was presenting.[45]

All who worked at Radio Hall in those years remember the social interactions of what McCarty called "the WHA Family," from the mandatory half-hour "coffee break" in the musty Radio Hall basement that interrupted the workday at 10:00 each morning to alcohol-centered gatherings that sometimes brought them together again in the late afternoon. In those days, Tom Clark was assistant program director. He remembered fondly the fellowship he found at Radio Hall but conceded that the organization "had no vision other than the status quo." Karl Schmidt described a colleague as "program director of a schedule that allowed little direction." From the mural that dominated the main lobby to the historical plaque outside that proclaimed WHA "the oldest station in the nation," Radio Hall meant comfort and tradition for those who worked there and for thousands of loyal listeners across the state.[46]

4

A STATE TELEVISION NETWORK?

1952–1965

Before H. B. McCarty and Harold Engel had completed their FM network, a new opportunity dropped in their laps. Intellectually they understood the importance of television for educational broadcasting, but emotionally they did not have the passion they had felt twenty years earlier for their first love, radio. Now, when they were ready to relax and enjoy the fruits of their labor, they were asked join a national effort to duplicate in television much of what they had already done in radio. They did join that national effort, but they did not change the name of the State Radio Council to include the visual medium, nor did they move their offices from Radio Hall to the television center emerging across the street. And, unlike what happened at other universities, they did not cannibalize radio to nourish television.

Still an important national leader in educational broadcasting, McCarty served with four others on a committee to define "the social role" of educational television. Convened in 1953, McCarty's committee called television "the most powerful communication mechanism ever." Members said educational TV offered the opportunity to strengthen the values of America's free society, to develop a well-informed and responsible citizenry, to increase the range of cultural choices, and to promote free and critical inquiry into problems of general concern. The committee predicted that educational TV "will point out to viewers what might be, as well as what is."[1]

It was McCarty and Engel's vision, creativity, political skills, and total

dedication that created a radio system unique to Wisconsin and its university. It was Henry Ford's tax lawyers who held the key to the development
of educational television. When Ford died in 1947, he left the bulk of the
Ford Motor Company to the family's foundation, creating a funding source
larger than the budgets of most national governments. The foundation
chose to fund, among other things, a national system of "educational television."[2] The Ford Foundation approached that challenge on three levels.
First, it organized and paid for a lobbying effort to reserve frequencies for
noncommercial educational television. Next, it provided matching funds
to construct stations. Finally, it established an Educational Television and
Radio Center in Ann Arbor, Michigan, to exchange programming among
the stations it helped build.[3]

The experience of AM radio confirmed the predictions of those who
said that profit-driven broadcasters would prevail over noncommercial
broadcasters in competition for frequencies, if some were not reserved
for noncommercial use. The fight for reserved television channels took
place between 1948 and 1951 and ended with 10 percent of the available
television frequencies designated for noncommercial use. In Wisconsin,
the Federal Communications Commission reserved twelve channels for
education, enough to provide coverage of the entire state.

With the channels reserved, Ford needed to make certain that noncommercial interests applied for and built the stations. To spur immediate
action in the slow-moving world of education, the foundation targeted
fifteen "educational centers," offering a grant of one hundred thousand
dollars to institutions that would match it. Not surprisingly, Ford targeted the University of Wisconsin. McCarty and Engel were first in line
to ask for one hundred thousand dollars to build a station to operate on
Channel 21 in Madison.[4] The UW regents provided the matching funds.
The State Radio Council, the independent state agency administered by
the university, got the license. The regents justified the "experimental"
station as a place to train students for professional careers and to broadcast instructional programming to Madison-area schools and to students
on campus. The idealistic language of the national committee on which
McCarty served was not cited in the decision to build a narrowly defined
"experimental" station.

WHA-TV took to the air May 3, 1954, as the nation's third noncommer-

cial TV station.[5] McCarty wrote that radio had justified the hopes and faith
of the pioneers who advocated it thirty-five years before, adding, "Now
comes television," with "even greater promise as an educational tool."[6]

WHA-TV's transmitter sat in Radio Hall and broadcast from the
WHA-FM tower behind the former heating plant, but a decaying campus
building at 600 North Park provided studio and office space. Built originally for engineering, 600 North Park had gained notoriety as the scene
of Professor Harry Harlowe's controversial psychological experiments
that deprived young monkeys of maternal affection. During Harlowe's
reign, poor penmanship gave "600 N. Park" the nickname GooN Park.
Other than its location directly across Observatory Drive from Radio Hall,
GooN Park had little to recommend itself as a television facility. The floors
sloped downward toward Lake Mendota just enough to challenge anyone
attempting to push one of two bulky, tube-laden cameras in a straight line.
The challenge was particularly great when the camera operator was an
inexperienced student. The pillars in the studio, while necessary to hold up

WHA-TV staff fought sloping floors and disruptive posts to produce live drama at
"GOON" Park. IMAGE COURTESY OF THE UW-MADISON ARCHIVES, #S14669

the creaky building, provided another challenge to the student crews operating the equipment. WHA-TV remained at GooN Park for eleven years. As
a wrecking ball took down the building, WHA-TV moved to a temporary
facility west of campus on University Avenue.[7] Staff hoped their permanent new home would rise at the GooN Park location overlooking Lake
Mendota, but the university gave that site to the Helen C. White Library.
Broadcasting ended up at Vilas Hall, built a couple of blocks to the south.

While stuck in far from ideal facilities, the GooN Park crew turned out
an impressive schedule of live programs each evening between 7:30 and
9:30 p.m. Monday through Friday. WHA-TV's production style was essentially radio with pictures, in part because McCarty's radio staff produced
much of the programming and because the limited facilities permitted
little more than in-studio "talking heads." Aline Hazard's *Homemakers'
Program* spun off a weekly television version, along with farm features and
music appreciation programs familiar to State Station listeners. Spanish-
and German-language instruction once done on radio now appeared on
WHA-TV. A journalism professor did a newscast, while Roy Vogelman
presided over a news/interview program in which Madison journalists and
students quizzed newsmakers. Another panel of students would ask faculty
members about their area of expertise in *Quiz the Professor.* UW folklorist
Robert Gard did a program on Wisconsin culture, while radio's Ken Ohst
conducted a weekly sports show for the UW athletic department.[8] All these
programs were live, in living black and white, and interspersed with kinescopes (films) of programs from other educational stations.

The broadcast schedule began each evening at 7:30 with *The Friendly
Giant*, a fifteen-minute program for children with WHA Radio's Bob
Homme in the title role, assisted by his *Bandwagon* sidekick Ken Ohst.
Ohst provided voices and manipulated the puppet heads for Jerome the
Giraffe and Rusty the Rooster. Homme had graduated from UW with an
economics degree about a year before WHA-TV hit the air, and he began
thinking about how he could transition his radio sensibility to television.
He drove to Chicago to watch rehearsals for the pioneering TV show *Dave
Garraway at Large*, another radio personality transitioning to television.
Thinking about a children's program while driving back to Madison, he
had a revelation: "If the sets and props are miniatures, then I can be a

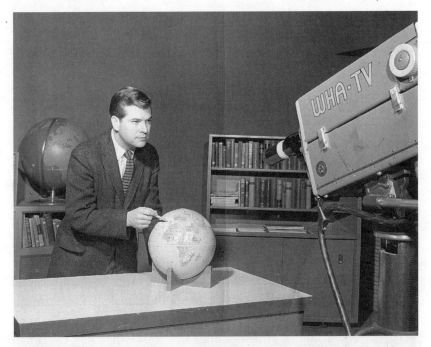

Geography professor Ric Johnson used only a desktop globe to explore *The Big Wide World* from the WHA-TV studio in 1958. WISCONSIN PUBLIC TELEVISION

giant."[9] Aware that giants could scare kids, he decided a friendly version of a scary thing might really appeal to them, and so came the concept of a giant named Friendly. While still working full-time for radio, Homme received approval to develop his television idea. His experiments revealed that he could not execute the concept as easily as he had thought. When shot from the normal camera angle, a grown man standing in a miniature set looked like a grown man standing in a miniature set. To create the desired illusion, the camera would need to shoot from the level of the miniature set, so the human figure could tower above it. He found, too, that his simple puppets did not look lifelike on TV. Jerome the Giraffe was not ready for prime time when the show began its run.

The Friendly Giant opened with the camera panning a miniature scene as Homme talked. Suddenly, the camera focused on a giant boot, and Homme said, "Look up. Look *way* up." The camera found Homme's friendly face and the program was under way. As on radio's *Bandwagon*

Radio's Bob Homme became the "Friendly Giant" on WHA-TV as soon as the station went on the air in 1954. IMAGE COURTESY OF THE UW-MADISON ARCHIVES, #S14761

Correspondence School, Homme and Ohst worked without scripts on the TV show to sound conversational. They played off each other, improvised comedy, told stories, and sang songs.[10]

Within a year, the Ford Foundation's Educational Television and Radio Center (ETRC) began national distribution of *The Friendly Giant,* its first program for children. From its headquarters in Ann Arbor, Michigan, the center distributed kinescopes of television programs to the handful of pioneering educational stations around the country. WGBH Boston, WTTW Chicago, WQED Pittsburgh, and KQED San Francisco were among the fifteen stations that broadcast the *Giant*. In 1956, an Ohio State Award for educational radio and television programs, then educational broadcasting's highest honor, called the program a "refreshing and intelligent approach to the entertainment needs of the pre-school child." It added that Homme and Ohst met those needs "positively without in any way patronizing or exploiting the child."[11]

A programming executive at the Canadian Broadcasting Corporation evaluated the program for his network and concluded that *"The Friendly Giant* is an artistic performance of rare quality." The executive went on to specify how the program related to academic studies on child development. The evaluator particularly liked the structure of the program, which gradually slowed down to create a "bedtime" experience by its 7:45 p.m. conclusion.[12] Indeed, the CBC respected *The Friendly Giant* so much that it hired Homme and took the production to Canada in 1958. Ken Ohst stayed behind in Madison. The ETRC continued to distribute the Canadian production to American educational TV stations, including WHA-TV, for another ten years. The program continued in Canada for fifteen years after that.

In addition to its two-hour evening schedule, WHA-TV came back on the air one afternoon each week to broadcast "experimental" instructional programs to Madison-area schools. Mostly, these programs were TV versions of radio's *School of the Air* productions. They included Fannie Steve's rhythm and games program, *Ranger Mac*, and *Let's Draw*. Other offerings were the *Young Experimenters* science program produced by a UW–Whitewater professor, and *Let's Have Fun*, produced by speech department graduate students.[13] Classrooms within fifteen miles of Madison could view the programs. They fell within the limited coverage area for the "experimental" station.

Channel 21 in Madison was only one of the twelve frequencies in Wisconsin the FCC had designated for educational use. McCarty's State Radio Council wanted the other eleven to build a television network paralleling its radio network.[14] Television proved a far greater challenge, however, because it had not had thirty years to evolve from a quiet physics experiment as radio had. Each new FM station constituted only a relatively small expansion, because it could use the programming and staff already in the budget for the original two AM stations. Engel and the Depression-mentality engineers who built the FM network did everything as cheaply as possible. They bought war-surplus equipment and repurposed whatever they could find. For example, an engine from a Model T Ford powered the emergency generator in Radio Hall. Staff did not discard equipment ready for replacement; they repurposed it to lower priority uses, continually expanding the size—if not the reliability—of the technical operation. Those who demanded the highest professional standards might shake their heads at the budget-stretching ways of the State Stations' engineers, but without their frugality the state radio network might not have happened.

Television was another matter. The State Radio Council proposed to build twelve television stations. That expense alone was far beyond the incremental costs of the individual radio stations. Moreover, these stations required higher technical standards, particularly if the Ford Foundation were going to pay part of the cost and retain part ownership of the equipment. The cheap "off-the-air" pickup system that relayed radio programming across the state would not work for TV. The new medium would require a real interconnection system. While "experimental," WHA-TV was providing a limited program service in Madison; a statewide TV service would require much more in quantity and quality to justify the investment in the facilities proposed. Engel, who had proved effective in convincing individual legislators to support improvements in the radio system, now needed to persuade the entire legislature to build a more expensive system almost from scratch.

Moreover, the political environment had changed. The state radio stations began in the Progressive Era, an optimistic time when people believed that they could address common problems through government action. Television arrived in the post–World War II period, when fear of communism and authoritarian enemies overwhelmed progressive optimism about the power of education and structural reform to strengthen democracy and improve the quality of life. In the first Republican primary after the war, Joseph McCarthy had defeated the embodiment of Wisconsin progressivism, "young" Bob La Follette. McCarthy came to embody the political mood in the state and the country that sought to eliminate subversive, anti-American influences in all areas of government and society. This "McCarthyism" led a special congressional committee to investigate the power of not-for-profit foundations, foremost among them the Ford Foundation. Staff at Ford believed the congressional investigators took a particular interest in the foundation's support for noncommercial television.[15]

While Ford provided the national momentum for educational television, Engel claimed that support for a Wisconsin television network sprang up "spontaneously" at the February 1952 Farm and Home Week conference conducted by the College of Agriculture. If Engel himself lit the match that combusted into a movement for educational television, he never admitted it. Whatever the spark, "practical farmers and farm leaders," not academics, convened a meeting the following May from which emerged the Wisconsin Citizens Committee for Educational Television. Milo Swanton, a farm leader, chaired the committee.[16] He pulled in the support of a wide variety of organizations ranging from the state Parent Teacher Association and the American Association of University Women to farm cooperatives and labor unions. Swanton made his pitch for a state-wide network to the legislative council that September.[17] He said a state TV network would provide practical information to farmers and housewives (both urban and rural) and health and safety information to the general public. State vocational schools could serve industry with televised training for employees. The network would reach students from elementary to college levels with course materials related to the arts, sciences, nature study, history, and music.[18] He did not make the sweeping claims about advancing American democracy and culture McCarty's national committee had. Swanton argued a very narrow "educational" purpose to justify the state's investment. He asked the legislature to provide immediate funding for the first two stations in the system, one in Milwaukee and the other the upgraded WHA-TV in Madison. He urged that planning begin for ten additional stations across the state.

The legislature gave Swanton some of what he asked for in its 1953 session. It instructed the governor to apply for all twelve channels designated for education in Wisconsin on behalf of the State Radio Council. It also provided funds to the council to help support the university's experimental broadcasting on Channel 21 in Madison. The legislature did not, however, provide funding for building the first two stations, a full-power Channel 21 in Madison and Channel 10 in Milwaukee. Instead, it punted the decision to a statewide referendum in November 1954.[19] The referendum asked voters, "Shall the State of Wisconsin provide a tax-supported statewide noncommercial educational television network?" The question modified the phrasing of the original, which did not include the phrase "tax-supported."[20]

The Public Expenditure Survey of Wisconsin, an anti-tax organization, pulled together the Wisconsin Committee on State-Owned, Tax-Supported Television to oppose the referendum. It gathered a coalition of commercial broadcasters, Republican politicians, anti-tax groups, and newspaper editors.[21] Committee leaders said they were "impressed" with the concerns of those worried about giving the state too much power.[22] "Pro-America" groups around the state opposed all new taxes and additional forms of "state control" and took a particularly virulent stance against the whole idea of a state TV network.[23] The Committee on State-Owned, Tax-Supported Television said it did not want a statewide curriculum for schools, but argued that programming that merely supplemented local curricula was not worth the money.

The Catholic bishop of Green Bay urged a no vote because poorly funded Catholic schools "would have no choice" but to carry the programs and give the state too much control over what was taught.[24] An advertisement by a northeast Wisconsin business group asked, "Do you think that the teacher in Madison can do better than the one in the room?" It then pointed to a four-million-dollar price tag that "must come through increased taxes."[25] Another advertisement in Milwaukee asked, "Do you want your high taxes to go still higher? Do you want government-controlled propaganda in your living room?"[26]

Both sides took up the potential role for commercial broadcasting in education. The Wisconsin Committee on State-Owned, Tax-Supported Television asked for a poll of commercial operators about the number of requests they had received from schools for broadcast time and how many they had turned down, presumably to make the point that commercial broadcasters would respond if the schools only asked. Engel responded diplomatically by suggesting commercial broadcasters "strengthen their presentation" by specifying the total time each week they were prepared to commit to educational broadcasting and "how much money they would contribute to the planning, preparation and presentation of educational programs."[27]

Volunteers promoted a State Educational Television Network, but voters rejected it in November 1954. IMAGE COURTESY OF THE UW-MADISON ARCHIVES, #S14830

The 1954 State Republican Convention passed a resolution opposing the state network, but Republican Governor Walter Kohler wrote Swanton that the resolution was adopted hastily, without due consideration by the delegates and with only about 10 percent of them on the floor when the vote was taken. He assured Swanton that the resolution did not represent most Republicans' views, and certainly not his. Kohler reminded Swanton that the present state broadcast system was "created by Republican governors and legislators."[28] Nonetheless, he said he would abide by the decision of the people in the November referendum. He added, however, that he would oppose any legislative action to close the state radio network.[29]

A leader of the national movement for educational television warned its Wisconsin allies that they needed "big money fast" to counter the campaign against them.[30] A sympathetic state legislator told the Wisconsin Citizens Committee for Educational Television that it had to more aggressively make its case: "The interests which oppose educational TV are hard at work telling their story." The state would not get educational television, he predicted, if proponents did not tell their side of the story, which he felt most state residents would support.

Looking back at his twenty-five years working with McCarty and Engel, Karl Schmidt believes they had lost the drive to carry out a hard fight.[31] Others contend that Engel and other educational TV advocates moved too quickly to initiate legislation without taking the time to properly plan their fight against highly motivated opponents. A week before the referendum, even the State Radio Council was getting cold feet. It passed a resolution saying WHA-TV demonstrated the success of educational television, but the resolution stopped short of saying the state should pay for a statewide network, in effect, conceding the election.[32]

The pessimism proved prophetic. The referendum failed by two to one, winning only in Dane County.[33] It lost decisively in Milwaukee County, where the local technical college had applied for Channel 10 in competition with the State Radio Council. The two had been unable to work out a compromise.[34] The results of losing the referendum were that Milwaukee secured a local educational TV station, Madison would continue to use the low-power "experimental" Channel 21, and a group in Duluth took the channel allocated for northwest Wisconsin. The rest of the state remained without educational television for another fifteen years.

In WHA-TV's early years, another genial giant, William Harley, presided over GooN Park, the same William Harley who had served as radio manager during McCarty's World War II absence and as radio program director (with little room for direction) in the postwar years. When television came along, McCarty named Harley, his younger protégé, TV program director, although everyone acknowledged that, ultimately, McCarty called the shots. Harley established an independent national reputation for himself with Ford Foundation–funded productions, including a nationally distributed television series, *Crisis in the City*. In 1962, Harley's success as a producer, his Midwest roots, and his sophisticated manner made him a politically acceptable choice to fill the newly created position of president of the National Association of Educational Broadcasters. This full-time assignment made him the national spokesman for educational television in Washington and leader of efforts that ultimately resulted in federal support for public broadcasting and the creation of the Public Broadcasting Service.

Lee Sherman Dreyfus filled the vacancy created when Harley departed in the early 1960s. Dreyfus's father was a Milwaukee radio personality and his mother, a civic leader. He received undergraduate, master's, and PhD degrees from UW–Madison before going off to Wayne State University in Detroit as a communications professor and manager of its educational radio station and educational television production center. He came back to Madison in 1963 as a professor in the speech department and "manager" of WHA-TV. It was the same year the station shed its "experimental" status by relocating its transmitter and increasing its tower height and power.

A highly engaging speaker, Dreyfus (known as LSD to his students and friends) delivered lectures about radio and television that packed classrooms and generated accolades from students. He became one of the most popular professors on campus. Student TV staff member Boris Frank remembers Dreyfus sitting on the main staircase of G00N Park holding court with a shifting audience of students. He would ask provocative questions and offer insightful, often witty critiques of the comments the crowd tossed at him. Frank recalls him as a great sounding board and critic.[35] The other half of Dreyfus's split appointment—manager of WHA-TV—proved more frustrating. Two years after arriving on campus, Dreyfus told university administrators he thought he had come to Madison to run the television operation. He said he would still like to do that, "but would need absolute clarity about responsibility, budgets, goals, and authority," which, of course, McCarty had never really relinquished.[36] The day after Dreyfus's first Christmas as TV manager, McCarty objected to the special holiday programming he had purchased. McCarty wrote that he had watched Channel 21 with "a gnawing sense of guilt if challenged to defend the costs."[37] A year later, Dreyfus protested that he needed eighteen hundred dollars for nine top classic films for a film history series rather than the twelve hundred dollars McCarty allocated for twenty-four "mediocre" ones.[38]

Earlier that year, Dreyfus had studied the commercial ratings for Channel 21, a measure McCarty would never consider relevant to the quality of service that either radio or television provided. In a curious choice of words, Dreyfus referred to men and women in the audience as "fathers and housewives." More presciently, he observed that "three children's programs that are not particularly good" had the highest ratings. He suggested that children made the best target for serving "alternative" audiences.[39]

While H. B. McCarty thwarted him as manager of WHA-TV, Lee Sherman Dreyfus thrived as a popular professor and ultimately as Governor of Wisconsin. IMAGE COURTESY OF THE UW-MADISON ARCHIVES, #S14721

Their strong personalities and differing philosophies defined the gulf between McCarty and Dreyfus. Above all a teacher, Dreyfus saw the university television station as a lab where students learned the art and science of television and as a place to produce instructional materials for campus use. Dreyfus's student-centered vision of educational television mirrored that of most universities. They saw television as part of a university's teaching and research mission. McCarty's vision, by contrast, went back to the Wisconsin Idea, which placed public service on a par with teaching and research in a public university. For McCarty, any training students received was incidental to the service radio and television provided to the public. Any opportunities to conduct research at the broadcast stations were similarly secondary to public service. It was that commitment to public service that set Wisconsin apart from other universities and made its broadcasting special.

University president Fred Harvey Harrington shared McCarty's desire to preserve the Wisconsin Idea in the face of the crush of baby boom students and government money for research that poured into the University of Wisconsin in the 1960s. Everyone still nodded to "the Wisconsin Idea," of course, but no faculty members bowed to it as deeply as they did to teaching and, above all, research. As part of a broader reorganization of the university to accommodate its growth, Harrington tried to restore public service to its rightful place at Wisconsin. He created a separate institution, University of Wisconsin Extension, to focus solely on public service and distance education, while the Madison campus continued to focus on teaching and research.[40] To head his creation, Harrington selected Don McNeil, one of his former students. McNeil was a young man with no experience in distance education, but he was smart, energetic, and dedicated to the Wisconsin Idea.

Harrington pulled into the new institution the second-class citizens and the campus misfits, contemporary versions of William Lighty. The Progressive-era General Extension division that Charles McCarthy founded in 1906 had diminished over time, but it formed the centerpiece of this new campus without walls called UW Extension. General Extension taught correspondence courses to students throughout the world and provided not-for-credit continuing education programs throughout the state. The massive Cooperative Extension operation from the College of Agriculture provided the greatest chunk of staff and budget for the new UW Extension. These were the county agents who helped residents deal with problems in agriculture, home economics, economic development, and natural resources. They ran the 4H programs in all seventy-two counties. Last and least, UW Extension gave radio and television a logical home in the new institution totally dedicated to the Wisconsin Idea of outreach to all residents of the state.

In the same July 1964 letter to a former colleague in which he had said, "Mac is still on his horse here and it ain't charging anymore, even at windmills," Karl Schmidt reported that "there's an exec vice president of the University who is riding him pretty hard."[41] He was referring to Robert Clodius, who found McCarty's independent ways frustrating as he tried to bridge the McCarty–Dreyfus gulf. About to become chancellor of the Madison campus under the impending reorganization, Robben Fleming wrote to Clodius with solutions to the problem that neither McCarty nor Dreyfus would work for the other.[42] Fleming said he did not see Dreyfus as "capable of running the whole show." Therefore, he recommended assigning Wisconsin's future governor (Dreyfus served one term as Wisconsin's governor from 1979 through 1983) to full-time teaching on the Madison campus. But, Fleming went on, "If Mac is put in charge of everything, he must be carefully directed (particularly to overcome his lack of interest in instructional media)."[43] President Harrington concluded the discussion by observing that on July 1, 1965, Mac would indeed have someone over him—the chancellor of Extension: "This in itself will change things; down to the present, Mac has reported only to the president or vice-president (really to no one)."[44]

5

UW EXTENSION

1965–1967

At 7:00 on a Sunday morning, a white-haired man with an upbeat manner and the unrelenting smile of a door-to-door salesman roused the new chancellor of UW Extension out of bed. Still in pajamas, Chancellor Don McNeil opened his door, and H. B. McCarty announced, "I just stopped by to say I am all aboard and ready to go to work for you."[1] By so audaciously introducing himself to his new boss, McCarty over-compensated for his unease at his new place in the university's bureaucracy after thirty-five years as a more or less free agent.

UW Extension provided a legitimate administrative home for radio and television, and some believe Harold Engel quietly advocated its inclusion.[2] For McCarty, however, the move meant a loss of independence, a loss of access to the university president's office, and, more disturbing perhaps, a potential change in mission for educational broadcasting. Until now, he had maneuvered on a campus that valued research and teaching with a faculty that saw little relevance for broadcasting to their primary interests. Radio and TV could do whatever McCarty wanted as long as they did not drain too many resources from activities deemed more important. The General Extension and Cooperative (Agricultural) Extension faculty pulled into the new UW Extension had few research or classroom teaching responsibilities. Their assignment was strictly outreach or public service, and broadcasting was the perfect tool for reaching broad audiences outside the classroom. They had every reason to assume that university administrators had put radio and television in UW Extension to use the medium for those purposes.

H. B. McCarty, left, and Harold Engel reluctantly gave up the reins of radio and television after more than 35 years. IMAGE COURTESY OF THE UW-MADISON ARCHIVES, #S05806

Two years after this forced marriage and forty years after joining WHA as a student announcer, McCarty stepped down. Engel left with him. McCarty's final communication as director of broadcasting embodied the same upbeat tone that had always characterized his leadership. He began with Charles Van Hise's oft-quoted statement that he would not rest content until the beneficent influences of the university were available to all residents of the state. He continued, "Van Hise and [William] Lighty had a wonderful dream. It must be modified, however, to embrace the greater opportunity for extending the university's 'beneficent influences' through radio, television, and the miracle of space satellite communication." He acknowledged that UW Extension provided the means to reach that goal.[3]

In the year before his departure, McCarty assisted McNeil in identifying a replacement. McNeil chose a former UW student who had risen through the ranks at commercial WTMJ-TV in Milwaukee, moved on to the Ford Foundation's Educational Television and Radio Center (ETRC) in Ann Arbor, and finally reached the top job at the educational television station in Los Angeles. To complement his high-level national experience, Jim Robertson had Wisconsin roots. Besides, he was a very affable person. Upon his arrival, Robertson declared his priorities for the state radio network and WHA-TV. It was a television-focused list that began with education for preschool children, the group Professor Dreyfus had identified as public TV's most promising target audience. The list had another Dreyfus priority, courses for campus-based university students. Then came courses for children in schools and for students enrolled in distance education. The list covered all varieties of instructional programming. Toward the end came programs for "housewives seeking intellectual stimulation" and, at the very end, for the "less well-defined and sometimes scarcely recognized requirements for continuing learning throughout life."[4] McCarty would have had no trouble defining the "requirements for continuing education throughout life" and would have placed it at or near the top of his list along with service to children in schools.

To compensate for his willingness to move down from the nation's second-largest media market to Madison and to accept, presumably, a commensurate pay cut, Robertson received time off to run a consulting business and to work from his second home in Door County.[5] Clearly, Robertson did not intend to match McCarty's seven-day-a-week commitment. Accordingly, Robertson delegated responsibility for radio to longtime staff member and *Chapter a Day* reader Karl Schmidt. To run television, he imported Ron Bornstein, a forceful young television producer from the University of Michigan. They had worked together at the ETRC in Ann Arbor. Bornstein had never managed a television station, but Robertson knew that he possessed the ambition, the intelligence, the political savvy, the charm, and, if necessary, the ruthlessness to do the job. Such skills and strengths would prove necessary as television fought to define its role within UW Extension.

Chancellor McNeil brought new respect for the second-class citizens thrown into his domain. Envisioning a campus without walls offering a full range of course work, he organized his disparate troops into academic departments and set up a system of faculty governance that more or less paralleled the structure of the Madison campus. UW Extension conferred academic rank and, in some cases, tenure on county agents and others who would never qualify under traditional academic standards. Broadcast staff members with master's degrees gained faculty rank in UW Extension. Broadcasting found its requests for upgraded equipment approved by the new chancellor, who saw these upgrades as central to the institution's mission.

With those upsides came some downsides. The newly empowered faculty of General Extension and, to a lesser extent, Cooperative Extension, saw themselves as distance educators, experts in their content areas and in the delivery of content to nontraditional students. They saw educational broadcasters as experts in neither content nor continuing education. With some reason, therefore, they believed they should determine the content of UW Extension's radio and television stations. They were, after all, newly respected content specialists and adult educators. They were the experts in what the public needed to know about their fields. Who-is-in-charge had long been a source of tension between the General Extension faculty and broadcasting. At the very beginning of WHA-TV, for example, the General Extension faculty declared they were responsible for "extending the university" and therefore had "major responsibility" for education via television.[6] Even though McNeil had bestowed faculty rank on broadcasters, the others in Extension still regarded them as little more than technical resources to help "legitimate" faculty members accomplish their educational goals.

Radio program director Jim Collins reflected the tension at the time in an anguished memo to radio director Karl Schmidt late in the summer of 1968. For years, the erudite Collins had trolled the University of Wisconsin catalogue for courses he thought would make interesting not-for-credit radio broadcasts. He looked for great lecturers who filled classrooms and might enchant radio listeners, teachers who could make people care about what Collins described as "off the beaten track" topics such as "Chinese Civilization," "Modern African Literature," and "European Social and Intellectual History." Now, he lamented, the chancellor wanted to offer courses for credit, by which Collins assumed he meant "fairly basic or survey courses." Moreover, the independent study department would grant the credit and, therefore, had to approve the courses. Piqued, Collins proposed that he simply step aside and let the independent study program administer *College of the Air.*[7]

A UW Extension dean wrote Robertson to say that "programs must be based on the knowledge and capacity" of the university, which is found in Extension departments. Therefore, he argued, "members of the departments must be directly involved in all programs of direct relevance to them" to avoid mediocrity and criticism.[8] Professor Bornstein would have none of that. He told faculty members from other units they could tell him how to do television when they let him tell them how to teach English or history or poultry science. To relieve the pressure, the broadcast division asked staff to compile and report the names of all UW Extension faculty who appeared or consulted on programs. Bornstein encouraged radio and television staff to tap the expertise of Extension faculty, but he insisted that they, the broadcast professionals, would make the decisions about program content. Ultimately, Bornstein had to accept a television programming advisory committee of UW Extension content specialists, but he made certain they served as advisors only.[9]

McNeil helped minimize the conflict by investing in a subcarrier technology embedded in FM broadcast signals that only specialized receivers could hear.[10] That embedded signal could carry material for targeted audiences, including students enrolled in distance learning credit courses. This subcarrier would free the main broadcast channel to carry informal education programs for the general audience. Collins and the listeners would have programs with which they were comfortable. The subcarrier system essentially gave UW Extension two statewide channels, one for radio staff to program for broad audiences and the other for content specialists to program with only technical help from radio staff. Television did not enjoy the same luxury.

Whatever turf battles the move to UW Extension engendered were more than offset by the vigor the new chancellor brought to the tradition-bound

UW Extension recruited
Ron Bornstein to reshape
television in 1967.
IMAGE COURTESY OF THE
UW-MADISON ARCHIVES,
#S14833

units he inherited. Urban unrest in Milwaukee in the summer of 1967
provided the impetus for an initiative out of the comfort zone for educational broadcasting and other traditional Extension units. Although less
destructive of life and property than uprisings in other major cities, the
demands heard from racial minorities in Milwaukee—and the demonstrations that dramatized them—shocked Milwaukee's white residents. Even
more shocked were Wisconsinites living outside the metropolitan area.
Governor Warren Knowles called on all government units to do what they
could to address the problems, but even before the start of that long, hot
summer, McNeil had set a project in motion.[11] All divisions of UW Extension would shift their traditional focus from rural to urban Wisconsin.
A revitalized Wisconsin Idea would help residents of Milwaukee's inner
core deal with the problems they faced and help residents throughout
Wisconsin understand life in the inner city. The boundaries of the city
would become the boundaries of the state. McNeil assigned television and
radio to educate Wisconsin's white population about "what it means to be

black and live on Milwaukee's north side." He handed Robertson forty-five thousand dollars in special funding to do it.[12]

Bornstein took personal charge of the project. He had assistance from a former colleague at the University of Michigan, Ralph Johnson. Johnson had just started as WHA radio's production manager. The two worked directly with McNeil to produce a series of radio and television programs called *The Inner Core: City within a City.* Johnson did the bulk of the reporting. He spent two days a week in Milwaukee over the better part of a year talking to people, recording interviews, and producing thirty weekly radio programs. He told about a furniture con operation that regularly exploited inner-city residents. He produced a portrait of a seventeen-year-old young man who had been in Milwaukee for only seven months "and has managed to get himself into more trouble than most of *us* can arrange in a lifetime."[13]

Johnson's enthusiasm mounted as he uncovered stories that shocked him as a white man and would surely shock other white people in the state. "At this point," he reported to his colleagues, "I feel quite sure that the inner core project is going to be one of the most significant things ever done in

A crew films a demonstration for the inner core documentary, *Pretty Soon Runs Out.*
WISCONSIN PUBLIC TELEVISION

educational broadcasting."[14] He was correct. However, his radio portraits received less recognition than the other two aspects of the project, a television documentary film, *Pretty Soon Runs Out*, and a five-evening series of live discussions broadcast on the public TV stations in Milwaukee and Madison and statewide on the FM network.

Robertson wanted to contract with award-winning documentarians Jack Willis and Fred Wardenburg to produce the television documentary. It would use most of the chancellor's forty-five thousand dollars.[15] The two filmmakers had each produced reports for national audiences on social issues, particularly racial ones, in the South and in northern urban centers. For the Milwaukee documentary, they focused on housing segregation, a problem Catholic priest James Groppi dramatized that summer by leading marchers across a bridge from Milwaukee's north side to the white ethnic enclaves of its south. Robertson ran into a glitch, however. State law required that the governor approve no-bid contracts, and Governor Knowles had doubts about this one. Willis and Wardenburg went to work on the documentary with the contract languishing on the governor's desk. Meanwhile, Robertson argued that no governor should have control over public broadcast programming.[16] Ultimately, the conciliatory Knowles did sign the contract, but with the proviso that "state administrators dealing with the inner core situation" would review the film before it was broadcast.[17] At the preview of the documentary, Knowles walked out in the middle. He explained that he now understood why Milwaukee officials were so unhappy with it.[18] *Pretty Soon Runs Out* won a regional Emmy Award, a first for any educational television station.[19]

The inner core project reached its climax the week of March 29, 1968, with a five-day barrage of programming. A pro bono publicity effort supported it. Ben Barkan, a Milwaukee public relations impresario, organized publicity and enlisted additional support from two Milwaukee corporations. Radio ran all thirty of Johnson's radio programs that week. Indeed, it devoted half of its daytime broadcast hours to the topic, including special material on the *Homemakers' Program*, music programs, and even on *School of the Air* broadcasts. WHA-TV and Channel 10 in Milwaukee devoted their entire evening schedules to live discussions of inner core issues. These discussions aired live throughout the state on the FM network, and even some commercial stations rebroadcast them. Channel 10 provided the studio space and a director. UW Extension staff based in Milwaukee helped round up participants and audience members. Chancellor McNeil personally moderated the evening discussions, which proved as lively as expected.[20] Conspicuously missing from the discussions were Milwaukee's mayor and any representative of the police department, who took a "no negotiation" stance toward the protesters. Those who did participate ranged "from militants and moderates to apathetics and extremists. Side by side, and face to face, people brought their anguish, frustration, and hope to the table."[21]

Reaction was intense and overwhelmingly positive. As the *New York Times* reported, "The day after it was televised, the Milwaukee Council, after months of delay, unexpectedly passed a strong open housing ordinance. . . . Many Negroes said they believed the television program had something to do with it."[22] Radio manager Karl Schmidt called it "WHA's finest hour," and told McNeil it could not have happened before radio and television moved to UW Extension.[23]

The manager of a Milwaukee commercial television station, however, called the project a "total disaster" and a "disservice to Milwaukee." Claiming that "program participants were angry" and did not accurately represent inner core people, he said "WHA's irresponsibility" would make dialogue more difficult in the future.[24] A UW Extension dean, who wanted faculty control of WHA content, complained, "Our departments were not only not involved in planning and development, they were not involved in the programs." He groused that outside experts were used in place of "our own highly qualified and able faculty."[25]

Not long after presiding over WHA's "finest hour," McNeil moved on to the presidency of the University of Maine, but Bornstein carried on his commitment to high-impact projects. In contrast to McCarty's frugality, Bornstein's WHA-TV sought major grants and contracts from government agencies and foundations for projects that aligned with the educational goals of those funders.[26] In 1968, Wisconsin's Commission on Aging, for example, provided seventy thousand dollars for two years of programming "of special interest to the aging."[27] The station said the programs would defy a culture that values youth. Citing a long list of academic studies on geriatrics, the station promised that *The Time of Our Lives* would not only present television programs of interest and help to its target audience, but would also seek to involve its viewers constructively in their communities.[28]

A much more challenging project targeted rural people with relatively limited education and skills. These people watched less educational television than their more urban and better-educated counterparts. The question was: Who would benefit more from educational television than those least inclined to watch it? The US Office of Education asked educational TV stations to propose projects designed to reach them.[29] In 1968, it said yes to WHA-TV's request for two hundred fifty thousand dollars for its *Rural Family Development* project, or *RFD*.[30] After three years, *RFD* evolved into a slightly different project called *360*, and that, in turn, transformed into *American Pie Forum*, an attempt to mix education with entertainment, the ultimate *Sesame Street* for adults. Unfortunately, rural adults proved a far more elusive target than urban kids.

RFD spent $708,000 of US Office of Education money over three years from 1970 through 1972 trying to serve the "undereducated, hard-to-reach adult who is not reached by more conventional adult education programs."[31] Weekly TV programs anchored the project, supplemented by radio segments, booklets, and pamphlets, a toll-free information line, and a group of home visitors. The booklets told viewers where to get help for social services, health, employment, legal issues, leisure, and recreation. Home visitors followed up with viewers recruited to watch the programs. A typical *RFD* program broadcast May 7, 1971, started with a feature on a Wisconsin county fair and went on to ask, "What does summer vacation mean to some school children in Mt. Horeb?" A nutritionist talked about calories and nutrition, and, not surprisingly, the UW Extension nutrition committee raised questions about its content.[32] In the fourth segment, an expert told viewers about "inexpensive ways for summer fun in Wisconsin."[33] Each segment offered a related bulletin.

In its fourth year, project director Boris Frank persuaded the US Office of Education to provide funds to extend the program to thirteen states served by the Central Educational Network.[34] Now called *360*, the series focused on "career education" primarily for those who had not finished high school. Sensitive to criticism of all the different labels the television staff had used to describe its target audience, Frank warned staff members not to call its audience "disadvantaged, underprivileged, undereducated, underemployed, [or] culturally deprived." Rather, he described 360 viewers as people "who do not have a chance to get the education they need or find a satisfying job."[35] A typical segment told the story of a forty-four-year-old black woman named Louise, who, after working fourteen years as a custodian, started her own business making hand-stitched clothing: "It was there all the time," she said in the script, "but I didn't know it because I had never taken the time to study myself and what I wanted to do."[36] One listener responded, "As a black woman, I find these segments stereotypic, destructive, and insulting."[37] Her complaint illustrated the challenge relatively affluent, better-educated white men faced trying to provide "education" to poor people, particularly minorities and women.

In addition, all people "who do not have a chance to get the education they need or find a satisfying job" are not the same. Subsequent research found that rural people in the northern United States did not relate to *RFD* and 360 because they perceived an orientation toward urban black people

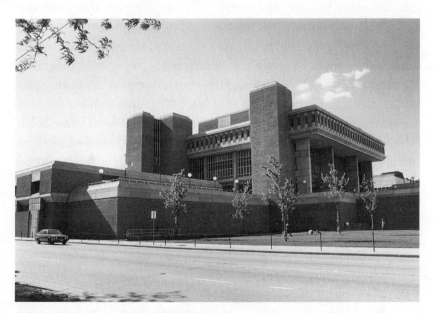

Radio and television moved to Vilas Hall in 1972. WISCONSIN PUBLIC RADIO

and southerners. After four years of funding *RFD* and *360*, the US Office of Education ended this experiment in adult education.

By 1974, WHA radio and television had relocated to the cavernous, state-of-the-art broadcast center in Vilas Hall. The main television studio was the size of a Hollywood sound stage, clearly designed for more than just small-scale instruction. About the same time, Bornstein promoted a brash young producer as WHA-TV station manager. Tony Tiano was a native of New Mexico but had a New York City style. A coworker described him as "rough around the edges," and "highly opinionated."[38] Tiano hired a New Mexico protégé, Joe Corrazzi, who enjoyed close ties to the Las Vegas entertainment industry. Together they took advantage of the spanking new Vilas Hall facility to add glitz and glamour to the remnants of the *RFD/360* project and gave it a new name, *American Pie Forum*.

Their two-million-dollar-a-year *American Pie* proposal promised to provide timely, relevant information about job opportunities for "adults for whom employment problems are aggravated by inadequate education, training, or knowledge of available resources."[39] The words of the song that closed the program captured the proposal's spirit:

Good-bye, dissatisfaction
Get your piece of the action
With the *American Pie Forum*
It's all for your job and you.[40]

The proposal promised a magazine/variety format starring TV personalities popular with a general viewing audience, among them Jackie Vernon, *Laugh In*'s Jud Strunk, Stanley Myron Handleman, Morey Amsterdam, Pearl Bailey, impressionist Charlie Callas, Betty Waldron from *The Flip Wilson Show*, and comedian Jackie Kahane.[41] As host, they hired David Canary, best known as Candy Canaday, a leading character on the legendary TV series *Bonanza*, and later the villainous Adam Chandler on the TV soap *All My Children*.[42] While Tiano never found the budgeted two million dollars, he pulled together nearly two hundred thousand dollars in federal revenue-sharing money to produce a no-holds-barred thirteen-program pilot series to demonstrate the concept and attract funding.

The cast bursts into song on *American Pie Forum*, WHA-TV's attempt to create a national *Sesame Street* for adults. IMAGE COURTESY OF THE UW-MADISON ARCHIVES, #S14668

The *American Pie Forum* was set around a pie factory (the American Pie Company) and a tavern nearby where American Pie workers hung out. As on *Sesame Street*, the action on *American Pie Forum* took place on a city street, this one outside the factory. The opening number had host David Canary come down the steps of an apartment building to a street scene full of American Pie employees. To the tune of "Pretty Baby," Canary sang: "Everybody wants a better job to get ahead someday," to which the crowd responded, "Applications. Aggravations."

Canary continued,

> Now I'd like to clue you in about the games you've got to play.
> (Applications. Aggravations!)
> If you want to land the right job
> You'll have to be sure
> To tell the man just what you can do.
> If you want to work, here's some advice
> I'll pass along to you, friend.
> Take a good interview,
> Yes, take a good interview.[43]

More songs and sketches followed in the tavern and on the street, each designed to educate while it amused, and probably doing neither as well as it might have had they chosen to do one or the other rather than both. Reviewers criticized the unrealistically upbeat nature of the messages, comparing them to Horatio Alger stories, inspiring perhaps, but essentially mythical. They questioned the white middle-class look of Madison residents pretending to work in the pie factory. In one segment, Pearl Bailey was the only nonwhite person in the scene. Reviewers enjoyed some of the comedy but did not see how most of it related to the lessons the program sought to teach. They felt the canned laughter and canned applause seemed, well, canned.[44] In short, the program wasn't yet where it needed to be. Nor was its funding. In a last-ditch effort to save the enterprise, Tiano asked the Internal Revenue Service for $895,000 to teach people about common problems on tax forms.[45] Perhaps *American Pie Forum* failed because it was poorly done, but more likely it failed because the station had taken on an impossible task. Adults are different from kids.

In any case, WHA-TV was unlikely to attempt another continuing education project of this magnitude. In 1984, for reasons that had nothing to do with radio or television, the University of Wisconsin system dispersed most UW Extension faculty back to Madison and other campuses. While still funding and coordinating distance education on all campuses, UW

Extension was left with direct operation of only two units, the county-based Cooperative Extension service and broadcasting. Gone was any dream of a credit-granting comprehensive open university without walls. For broadcasting, the change ended the program advisory committee and the pressure to vet programming through UW Extension faculty. This reorganization liberated WHA radio and television to fully embrace "public broadcasting," as described in the federal legislation that led to the creation of the Public Broadcasting Service and National Public Radio. The Public Broadcast Act passed in 1967, fifty years after Professor Terry's guests heard the "first broadcast" of 9XM.

The Public Broadcasting Act gave a new name to the concept of public service at the core of educational broadcasting in Wisconsin from the beginning. Neither Professor Terry nor Professor Lighty thought educational broadcasting meant formal instruction, college courses, or in-school programming. McCarty and Engel added those later. Terry had spoken of spreading "the university atmosphere" to the entire state, applying the values and approaches of a university across all types of programming aimed at the general public. President Glenn Frank said state residents needed "educative" material on a full array of topics to enlighten and stimulate. He wanted radio to "serve the interests of an informed public opinion" and "serve the recreational needs of the public." The Public Broadcasting Act of 1967 reflected these goals but also picked up the populist tone of Charles McCarthy's original *Wisconsin Idea* as it promised to empower the common person. The Act simply turned into national policy Wisconsin's original concept of public service broadcasting.[46]

PART 2

PUBLIC BROADCASTING

1967–2016

6

PUBLIC TELEVISION

1967–1987

On a Sunday evening in March 1968, President Lyndon Johnson surprised everyone, except his wife, Lady Bird, by announcing that he
would not run for reelection. The commercial networks went straight to
their journalists in Washington and New York for instant analysis. PBS
did not exist yet, but educational television carried the speech as part of a
Ford Foundation project to demonstrate how "non-commercial television,
when backed by adequate funds for programming, might produce superior
cultural and public affairs programs for a nationwide audience."[1] When the
president finished his announcement, the public television demonstration
project took its national audience to a bar on the east side of Madison, Wisconsin, to gather reactions from its customers. While commercial television
sought comments from credentialed experts, public television presented
the "alternative"—comments from the uncredentialed. In this demonstration, "public" broadcasting set out to illustrate its difference from "educational" broadcasting, and it chose to show that difference in Madison.

President Johnson proposed legislation in 1967 that created the Corporation for Public Broadcasting. CPB then created the Public Broadcasting
System and National Public Radio and provided federal support for noncommercial broadcasting. Critic Les Brown called public broadcasting "a
name without a concept," because it defined itself more in terms of what
it would not be than what it aspired to be.[2] The Carnegie Commission on
Educational Television, which came up with the term *public broadcasting*,
purposely left "instruction" out of its definition. Instead, it referred to

"educative" programming built on "quality" and "diversity." While the three commercial television networks entertained "mass" audiences, public broadcasting could serve an array of smaller, more discrete audiences.[3]

"America is geographically diverse, ethnically diverse, widely diverse in its interests," the commission proclaimed.[4] The commission, therefore, proposed a system built on "the bedrock of localism." This proposal disappointed those who wanted something like an American BBC that would educate public tastes, promote quality, and raise the level of public discourse. However, the possibility of federal money enticed the "Anglophiles" to at least give lip service to Carnegie's more democratic vision. Ultimately, BBC-style programming—as well as BBC-produced programming—would prevail on public television. In 1967, however, the Carnegie Commission had to satisfy southern conservatives suspicious of centralized media and new left-wing activists who sought "power to the people." It had to recommend a locally based public broadcasting system. The commission cited the New England town meetings where citizens once gathered to debate and decide issues confronting their communities. President Glenn Frank had used the same example when he set out his vision for WHA radio in the 1930s. The Carnegie Commission said problems urban areas faced demanded "the engagement of each individual citizen, who must be both informed and moved to act."[5] Educational broadcasting had long sought to inform citizens. Moving them to act echoed the original Wisconsin Idea but grew more directly from the atmosphere in the 1960s and Lyndon Johnson's Great Society initiatives of 1965 and 1966, which aimed to empower communities and eliminate poverty and racial injustice. From that point of view, going to citizens for comment on the president's decision not to seek reelection made sense. Of course, the Ford Foundation producers in New York could have gone anywhere to tap the bedrock of localism, but they chose Madison, Wisconsin, and WHA-TV. Under Bornstein, WHA-TV was more receptive than most stations to an activist vision of noncommercial broadcasting.

Determined to shake the WHA-TV staff out of what he perceived as lethargy, Bornstein sought a new manager with higher expectations and a "public television" vision. In 1968 he chose a dynamic graduate student he had worked with at the University of Michigan. Dick Lutz's graduate training had steeped him in theories about the role of media in democracy; his self-confidence and Bornstein's complete backing allowed him to pursue those theories wholeheartedly. While Bornstein wanted national recognition more than the implementation of a specific philosophy, Lutz wanted television to serve democracy, diversity, and localism. Lutz's uncompromising sense of mission, management style, and big voice terrified the staff Bornstein had hired him to transform.[6] Instead of a private office, Lutz placed his desk in WHA-TV's front lobby. He occupied the center of action, fully accessible to his staff, his staff fully accessible to him. Committed to establishing a more "professional" atmosphere, he promulgated rules and procedures. Nothing was too insignificant to warrant his intervention. He insisted, for example, on having the station vehicles washed every day so they always looked impeccable. A group of staff members complained to Chancellor McNeil that they felt "personally harassed [in] an effort to force them to resign" so the new management could hire its own people.[7] They perceived "a conscious effort to belittle what was done in the past at WHA-TV," which was destroying the morale of veteran employees.[8]

Whatever issues his management style might have raised, Lutz rebuilt WHA-TV on the bedrock of localism. The station was suddenly providing live coverage of city council meetings and local elections under the rubric the *Madison Vote-in*. The station's remote truck, traditionally used for UW sports broadcasts, found a new use covering legislative hearings on racial discrimination. The station invited Madison's anti-war congressman to hold televised public hearings on Vietnam in its own studio. It televised a "teach-in" with Senator Gaylord Nelson to launch what became the national observance of Earth Day in April 1970.[9] Lutz proposed that PBS schedule twenty-four hours of environmental programming to mark the first Earth Day, but the national network was less entrepreneurial and less committed to using television to promote civic action than the station manager at WHA-TV.[10] Madison's public television station reflected the 1960s activism of the community it served and the campus it called home.

In his most controversial initiative, Lutz persuaded the Ford Foundation to provide two hundred thousand dollars for a one-year demonstration project to produce a nightly newscast employing "disadvantaged youths" as "reporter trainees." The grant funded a "storefront studio" on Madison's not-yet-gentrified near East Side. In his request to the foundation, Lutz promised that this studio would allow "the community of the disadvantaged" to view the community at large "through new eyes, that we may see ourselves as we have failed to see before, a microscope finally reversed."[11]

To lead the effort and anchor *SIX30*, the nightly news show, Lutz chose an old pro journalist and political insider. Owen "Dick" Coyle had covered city hall for the "progressive" *Capital Times* newspaper before serving as assistant to Mayor Otto Festge, a prominent Democrat in the nonpartisan office. When Festge left office, Coyle went back to journalism at the refocused WHA-TV. He identified fifteen "trainee reporters," expecting attrition to reduce the number to a more manageable size. Attrition initially took only one.[12] While Coyle struggled to impose a minimum level of professionalism, Lutz defended the lack of it. He quoted PBS journalist Bill Moyers's pronouncement that "of all the myths of journalism, objectivity is the greatest." Lutz said the trainee reporters were first of all human beings with human concerns and human reactions to the events they reported.[13] From its beginning, *SIX30* sought to portray the community as the "disadvantaged" perceived it, not as professional journalists reported it.

Trouble started with the very first broadcast, although not because of any actions by the trainee journalists. One of the trainees interviewed radical student alderman Paul Soglin about his proposal to use methane and propane instead of gasoline to fuel city vehicles. Then professional journalist and anchor Dick Coyle took over and asked about rumors that Soglin would organize a recall campaign against Mayor Bill Dyke. Soglin replied that recalls could happen only after the official had a full year in office, and added, "It is obvious the city made a bad mistake in electing Dyke last year." Coyle closed the interview by saying, "OK, very good. Come back and visit again."[14]

Dyke was furious. He called the interview "an hour of assassination," but declined the station's invitation to respond on the next evening's program. Instead, he asked for a tape of the program to send to members of the state legislature. After reviewing the tape, Republican Assembly Speaker Harold Froelich said, "Quite frankly, from time to time I think the university needs censorship."[15] He and other legislators said they would block state building funds to upgrade the aged WHA-TV transmitter. When Lutz finally managed to meet with the mayor, he summarized Dyke's complaints in a memo to his files. Dyke, the memo said, felt Coyle,

WHA-TV station manager Dick Lutz announces the launch of *SIX30*, the controversial
news program that gave voice to young and disadvantaged reporters.
IMAGE COURTESY OF THE UW-MADISON ARCHIVES, #S14661

having worked for the *Capital Times* and Mayor Festge, was "out to get"
the Republican mayor. The program interviewed Soglin too often, Dyke
complained, even when he was not an expert on a topic. Dyke described the
storefront studio as "enemy territory" and refused to go there. He wouldn't
play on the other team's "home field." On a more philosophical level, the
mayor complained that a program like *SIX30* was inherently divisive and
that a state-owned entity, like WHA, should not cover local politics.[16]

The mayor was not the program's only critic. At the request of the UW
regents, the Extension University Committee examined letters and other
communications to the station and reported finding criticism representing the full range of concerns leveled against the Madison campus in the
late 1960s.[17] The "reporter trainees" were characterized as pro-Castro,
pro–Black Panther revolutionaries, unclean, unkempt, antisocial dissi-
dents, convicted criminals, and radicals. While "barely able to speak our
language," they propagandized an unsuspecting public in favor of homo-
sexuals, utter permissiveness, campus disturbances, and the downfall of
government.

Choosing progressive journalist Dick Coyle as host of *SIX30* helped poison relations with Republicans. IMAGE COURTESY OF THE UW-MADISON ARCHIVES, #S14658

Major figures in Madison's liberal establishment defended the project with lavish praise, although some were clearly torn. David Cronon, a prominent Madison historian who would soon become a highly respected dean, praised the concept of the program until it trod on his turf. During a strike by university teaching assistants, he wrote the station a letter, which read in part, "We have a right to assume that a news program on the university's television station will do full justice to the university's side of such a dispute. I do not think your program did so."[18]

The occupant of an endowed chair that honored WHA's first program director, Professor William Lighty, thought the program should continue if it labeled itself an opinion program rather than a news program. William Wedemeyer observed: "Rumors, opinions, half-truths go unchecked, unchallenged. Sources and biases are not probed. Representatives of the other side face a hostile environment."[19]

After reviewing the program and comments about it, the Extension University Committee recommended the administration give the program its full support. *SIX30*, the committee said, was doing exactly what it said it would do. "We would in no way recommend a curtailment of the opportunity for free expression by this group which presents the problems of society as they see them, and, in fact, it offers an excellent opportunity for other elements of society to hear firsthand some of the viewpoints to which they are not often exposed."[20] However, the issue was not that simple for Henry Ahlgren, who had replaced Don McNeil as Extension chancellor. The Ford Foundation had funded the storefront studio as a one-year demonstration project. At the end of the year, the two hundred thousand dollars would disappear. As a show of continued support, Ford offered one hundred thousand dollars for an additional year on the condition that UW Extension match that contribution and promise to take over full responsibility for the project in its third year. The chancellor was not willing to make that commitment, and *SIX30* announced it would end its run in June 1970.

Shortly after the announcement that the storefront studio would close, Lutz spoke to a group of Madison community leaders, his final address, it turned out, as station manager. Lutz did not mention *SIX30* in his talk, but he did state that he believed politics more than economics influenced the chancellor's decision not to save the storefront studio. Lutz quoted Edward R. Murrow's warning that "we must not walk in fear." He noted that Murrow's exposé on Senator Joseph McCarthy was not "objective" but was necessary to bring down a dangerous demagogue. A quote from former President Eisenhower's press secretary seemed an even more direct criticism of the chancellor's decision: "Pressure to force cancellation of a program and pressure after it to punish or intimidate . . . threatens the very existence of freedom of the press."[21]

Lutz was already scheduled to take a medical leave of absence from his position for reasons unrelated to *SIX30*. After leaving, he sent the WHA-TV staff a mock news release announcing a new program, *Pollyanna Views the News*. In the fake release, Lutz quotes himself as saying, "We think this new approach is certain to satisfy political critics who were unhappy with

SIX30's over-zealous reporting on matters needing correction in our society."[22] Lutz did not return to the station at the end of his leave.

The *SIX30* controversy effectively ended the possibility of building a state television network controlled by the university, according to Paul Norton, the legislative reference bureau staff member who worked on communications issues in the late 1960s.[23] Many legislators were already doubtful about public service broadcasting and about the university's involvement in it, and *SIX30* sealed their opposition. They might support "instructional" television, but they did not trust the university to control it. For much of its history, WHA had combined instruction with public service programming that sought to inform and inspire, enrich democracy, and raise the level of public debate. Harold Engel, however, had made instruction central in selling the state FM network as well as in his effort to sell a state television network. His pitch worked for the first and failed for the second. Now instruction was the only hope for a state television network in Wisconsin, even as public broadcasting blossomed nationally.

While Wisconsin had failed to build a television network in the 1950s, several southern states had seen an educational television network as a relatively inexpensive way to improve the sorry condition of their public schools and to equalize them without racially integrating them. Governors, not universities, controlled the politically appointed boards that directed these southern TV networks. They provided an alternative structure to McCarty's model of a university-state partnership, with the university as the dominant partner.

As soon as he arrived in Madison to succeed McCarty as director of broadcasting for both UW Extension and the State Radio Council in 1966, Jim Robertson proposed a strategy to build a state TV network. Robertson recognized that antagonism toward the university—and in some cases personal resentment toward his predecessor—would make the radio arrangement unacceptable for television. Understandably, the Department of Public Instruction could only support a TV network that emphasized school programming. The state university system, which boomed in the 1960s, opposed any more power or influence for its institutional rival, the University of Wisconsin. Widely read newspaper columnist John Wyngaard claimed that pursuing a state public TV network was merely rising to the "bait" of federal money. More fundamentally, he criticized the enterprise because "professional educators have not made a convincing case" that kids learn anything from watching television in the classroom. His column quoted an unnamed legislator who called WHA in Madison and WMVS in Milwaukee "educationally owned" television, not "educational" television. In fact, he said, they are thinly disguised competitors for commercial television. [24]

In his "informal initial comments" about building a state network, Robertson suggested turning to the Coordinating Council on Higher Education (CCHE), an agency established to reconcile the competing ambitions of the university, the state universities, and the state technical colleges. He would place the State Radio Council under the CCHE and change its name to the State Broadcasting Council. He would add more K-12 representatives and citizen members to the council and charge it with planning a television network to match the state radio network. Robertson proposed that UW Extension "furnish" its director of radio and television (Robertson) to carry out the work of the council in a dual appointment like the one McCarty had enjoyed with the university and the State Radio Council. [25] State money through UW Extension and the State Broadcasting Council would fund the joint enterprise, along with federal grants.

Democratic Assemblyman (and future Congressman) David Obey introduced legislation similar to what Robertson proposed. The state legislature created the Educational Communications Board (ECB) under the CCHE to replace the State Radio Council and to plan the state television network. Obey's legislation provided a little over a million dollars in state money to attract a federal match. Republican legislators proposed that the network carry "strictly construed instructional-credit course related programs." They said specifically that programming "should not be public service related," but that a future legislature might give approval for an expanded mission. [26] While Obey's bill passed without the Republicans' proposed amendment, the future congressman sold his legislation on the value of a state educational TV network for "instruction," K-12, vocational, independent study, continuing professional education, and even inter-campus exchange of instructional resources. [27] Only one rationale, "educational opportunity for all citizens of the state," left open the possibility of public or public service programming outside a narrow instructional mission. Governor Knowles signed the bill January 23, 1968. [28]

Governor Knowles endorsed placing all educational television (including WHA-TV in Madison and WMVS in Milwaukee) under ECB's control,
effectively ending any university role in public television. His structure
would parallel the one southern states used focusing on K-12 instruction.
Any doubt about ECB's intentions to break its historic tie to the university
disappeared when it hired the head of Georgia's educational television
network to head its operations rather than jointly appointing UW Extension's Robertson.

With the joint appointment model dead, Robertson resigned. He left
Bornstein to fend off the ECB takeover of WHA-TV the governor had proposed. Bornstein argued against a single organization controlling all of
educational television, particularly if it were a politically sensitive board
directly beholden to the governor and legislature.[29] Bornstein noted that
Milwaukee, Madison, and Duluth/Superior already had local public TV
stations. He suggested the ECB facilitate a "confederation of local stations"
with new stations in Green Bay, Eau Claire, La Crosse, and central Wisconsin, where he found an ally in Lee Dreyfus, by then president of Stevens
Point State University.[30] Bornstein emphasized the importance of local
service. The Federal Communications Commission licensed stations to
serve their communities, he pointed out, and the Corporation for Public
Broadcasting provided "community service grants." Federal policy wanted
broadcasters to serve specific communities, not entire states.[31]

Whatever the logic of Bornstein's argument, university president Fred
Harvey Harrington did not support it. He did agree, however, that the
ECB should not take over the license—and thus control—of WHA-TV.
Bornstein claimed, "The motivation for this license transfer was political,
stemming from conservative reaction to WHA-TV local programming—
specifically a Ford Foundation–sponsored series, SIX30."[32] Even though
UW Extension had declined to take over funding for SIX30 when Ford
Foundation support ended, Bornstein contended that the university remained better insulated from outside pressure than any state agency. "The
ECB's demonstrated reaction to irresponsible political pressure indicates
its political vulnerability, as well as its posture to avoid station involvement in controversial issues. Contrary to the ECB's attitude, the FCC urges
broadcast licensees to involve themselves in such matters," he wrote.[33]

Bornstein refused to lose on this issue and he didn't. He rallied not

The legislature funded the Educational Communications Board's state television network specifically for instruction, particularly for kids in K-12 classrooms. IMAGE COURTESY OF THE UW-MADISON ARCHIVES, #S14663

only Harrington and other university leaders but also Madison community leaders who comprised an organization he created called Community Council for Public Television, which eventually evolved into the Friends of WHA-TV. The ECB voted to take over the WHA-TV license, but the UW regents voted to retain it, and neither the governor nor the legislature chose to press the issue. WHA-TV and Channels 10 and 36 in Milwaukee retained their licenses and their independence, while the ECB would build and operate new stations in Green Bay, Wausau, La Crosse, Menomonie, and Park Falls and would interconnect all the stations to share programming. The ECB, however, seemed to regard its failure to take control of WHA and WMVS as a temporary setback. Had the agency accepted its role as a regional broadcaster, it would have placed its headquarters in the region it served, most likely in Green Bay. Instead, the ECB debated placing its headquarters in Milwaukee at Channel 10 or in Madison's proposed communication building, Vilas Hall. Either location might serve as the production center for a statewide network. Indeed, the university had designed the Vilas Hall radio and television facilities for that purpose when it assumed that UW Extension would lead the statewide television network as it had the statewide radio network.

Soon after arriving in Madison from Georgia, the ECB's new executive director, Lee Frank, asked the state building commission to halt construction of Vilas Hall until the ECB could review it. Privately, university officials conceded that the planned broadcast facility and equipment for Vilas Hall went far beyond what a local Madison station would need.[34] Publicly, university officials said they needed such radio and television facilities to attract national production contracts, irrespective of ECB's needs. Frank relented a month later, saying the television facility was necessary to produce programming for the ECB, although the radio facilities were far more lavish than anything ECB would need. Vilas Hall went ahead as a university building, with the proviso that ECB have facilities there to feed its network. Two control units sat side-by-side in the new Vilas Hall control room, one for WHA-TV and the other for the five ECB-owned television stations. Most of the time both units were sending the same programming, but they symbolized the independence of WHA-TV. ECB demonstrated its independence by hiring its own staff for in-school instructional programming, dropping the name *School of the Air* and making redundant the UW Extension staff members who had produced the programs since 1931.

For the next decade, ECB and WHA-TV engaged in a continuing cold war. Both chafed at the arrangement that left WHA separate from the state network and the ECB dependent on WHA for programming and facilities, but each did what it needed to do to make it work. Indeed, the years after the ECB's establishment may have been the best in the history of educational television in Wisconsin.[35] ECB built five TV transmitters, providing the long-sought statewide coverage for educational television. UW Extension funded WHA-TV generously, and viewers provided an additional million dollars a year through the Friends of WHA-TV. Those funds generated additional federal money from the Corporation for Public Broadcasting through a matching formula. State support for the five ECB transmitters elsewhere in the state and for instructional programming generated a similar matching grant from CPB, money the ECB could spend on public service programming. For one of the few times in its history, money did not present a major challenge for public television in Wisconsin.

Year after year, WHA-TV chalked up the highest ratings of any public television station in the country. It reached 49 percent of Dane County's population each week.[36] The station took advantage of its University of Wisconsin affiliation to broadcast Wisconsin sports, which accounted for some of WHA-TV's popularity, but generally WHA-TV tried to keep its distance from the UW–Madison campus and to embrace the Madison community. Program director Larry Dickerson recalls deflecting requests for airtime from faculty members. He enjoyed Bornstein's full backing to do so.[37] Bornstein declared he did not want to be the "panel channel," and Dickerson joked that educators thought a nun in front of a blackboard constituted exciting educational television.

Producer Tom Weston tapped the classic film archives of the State

Director of WHA-TV Ron Bornstein wanted to shed the station's image as the "panel channel," and dispatched programs such as this journalist roundtable. WISCONSIN PUBLIC TELEVISION

Historical Society for broadcast under the rubric of "education." A popular culture scholar, Weston touted WHA-TV's educational mission and
small-market status to acquire old television series such as *Burns and Allen,
The Twilight Zone,* and *Alfred Hitchcock Presents* for broadcast Saturday
mornings when commercial stations aired cartoons for kids. *Madison
Magazine* went after these choices in a major article titled "Commercial
Television Is Casting Wary Eyes on WHA."[38] It quoted a commercial station
manager saying he did not mind competition, but he did mind "somebody
tying one arm behind my back before the fight," referring to WHA-TV's
tax subsidy, which the article said came to three million dollars a year.[39]
Bornstein disputed that number, but he could not deny that WHA-TV
received 41 percent of its budget from tax sources, the same percentage as
it received from viewers.[40] The commercial station manager said "local"
programming on WHA-TV should cover the arts, not Badger sports, which
competed with his station for sponsor dollars.

Of course, public TV tackled important issues, too. Both WHA-TV and
the ECB took on significant projects during these halcyon days. Without
the ability to produce its own programs, the ECB contracted with WHA-TV
and independent producers for content for the state network. Glenn Silber, an independent producer in Madison, made two powerful documentaries for the ECB. *The War at Home* provided sympathetic treatment of
the anti-war demonstrations that rocked Madison in the Vietnam era.
The program earned a nomination for the Best Documentary Academy
Award in 1979 and remains a classic portrayal of the tensions in the anti-
war movement.[41] ECB also commissioned the less successful, but no less
controversial *An American Ism: Joe McCarthy,* which followed the career
of the Wisconsin senator from his boyhood to national notoriety. Ironi-
cally, these were exactly the kind of controversial projects Bornstein had
argued during the license transfer debate that a state agency would avoid
and a university station would embrace. The ECB program director, Byron
Knight, coproduced with a German broadcaster less controversial, but
well-received documentaries on composers Bach, Beethoven, Mozart, and
Haydn with funding from the Aid Association for Lutherans. Peter Ustinov
hosted the programs and PBS distributed them.

In the late 1970s, the program directors of WHA-TV and the ECB put
aside their institutional rivalry and pooled their resources to produce a

The ECB commissioned the controversial documentary, *The War at Home,* which garnered both criticism and acclaim for its sympathetic treatment of the anti-war movement.
IMAGE COURTESY OF THE UW-MADISON ARCHIVES, #S14835

showcase magazine program patterned loosely on commercial television's *60 Minutes.* Aside from its focus on the state, *Wisconsin Magazine* differed from *60 Minutes* in one important way. It would not "look for trouble" as the muckraking CBS program did in exposing crooks and corruption.[42] In fact, *Wisconsin Magazine* devoted relatively little time to public issues. It generated acclaim mostly for its human-interest features, slices of Wisconsin life, and explorations of Wisconsin history. All involved in *Wisconsin Magazine,* including host Dave Iverson, gave credit to the visual and audio artists who teamed up with the producers and reporters to tell the stories. WHA-TV program director Larry Dickerson called it television at its best because of how well it used pictures to tell stories.[43] Apparently, the Corporation for Public Broadcasting felt the same way, declaring *Wisconsin Magazine* the best local program in the country in 1984.[44]

Having achieved that recognition for their showcase series, the ECB and WHA-TV did something unheard of in commercial television. They killed *Wisconsin Magazine*. The reason says a lot about the culture of WHA-TV. Producers drove programming decisions, and *Wisconsin Magazine* absorbed resources at the expense of other activities that had more interest for the current producers. Some wanted more issue-oriented news and public affairs. The lower-budget *Wisconsin Week* let them pursue that interest while leaving resources to produce a handful of documentaries with high production values.[45] Some of those savings funded Dave Iverson's *Uncommon Places*. The series about the architecture of Frank Lloyd Wright earned high acclaim when it was shown on PBS in 1985, at least a decade before Ken Burns took on the same topic. As had been the case with *Wisconsin Magazine*, total collaboration between "the producers and the shooters" provided the extra edge of quality that distinguished this series and other major WHA-TV productions.[46]

Around 1980, some projects took WHA-TV producers and shooters far from Wisconsin to pursue their interests. The veteran head of the public affairs department, Carol Cotter, went to the southwest United States to examine the *Winds of Change* among Native Americans in that region. Another crew went to Vietnam, a decade after Americans stopped paying attention to the country that had played so large a role in US foreign and domestic life. Italy, Germany, Paris, and London highlighted the itinerary of another production team in their film *Search for the Violin*. This production studied the history, art, and, above all, physics of great violins based on the work of a retired UW–Madison physicist. The National Science Foundation liked the idea, but would not fund the project unless the PBS series *Nova* signed on as a partner.[47] *Nova* did, and the project went ahead. None of these documentaries related directly to Wisconsin, but each attracted national attention to the production capacity of WHA-TV.

During the post-Lutz decade, the bedrock of localism at WHA-TV and the ECB clashed with the urge to produce national-caliber programs on topics of interest to individual producers and managers. Perhaps Les Brown was correct to call public broadcasting a name without a concept, defined more by what it was not than by what it was. It was not education or instruction. It did not reflect what educators thought students or the public needed to learn. Rather, it reflected what creative producers wanted to produce or what managers felt the community wanted or needed. Born-stein sought to remain in the university and retain state support, but he used the privately incorporated friends group to take on the characteris-tics of a community station. He started with the Community Council for Public Television in 1968, which later became the Friends of WHA-TV. Each served as a counterweight to the university, the ECB, and the state government.

With active citizen leadership, the Friends advocated for the station. They had proved instrumental in keeping WHA-TV from the control of the ECB. They provided community input to station management and con-ducted fund-raising events around the city, some successful, others disas-trous. The most notable event, marking WHA-TV's twentieth anniversary, brought twenty-five thousand people to the Dane County Coliseum May 11, 1974, but the fund-raiser lost money when featured entertainer Jerry Lewis pulled out at the last minute because of poor ticket sales.

That same year, WHA-TV invited the Friends to use its airwaves to conduct an on-air membership drive, disguised as "Awareness Week"—a week that lasted seventeen days. Volunteers took to the air to ask for con-tributions. In general, the pitches for support were low key and nervously amateurish.[48] Indeed, the volunteers made their pleas so politely that casual viewers might not have realized they were supposed to pledge ten dollars to join, but four hundred new members did.[49] A second attempt later that year sought to entertain viewers with comedy sketches and celeb-rity interviews. The volunteers learned that "dull raises money, clever doesn't."[50] Two years later, the support group organized its first on-air auction, one of the largest volunteer events in the city. A forty-minute thunderstorm took WHA-TV off the air during the auction's first night, but the event proved a great success. While a strict cost-income analysis might show little monetary profit from the auction, WHA staff regarded the com-munity involvement as worth the effort.[51] By the early 1980s, the Friends organization provided as much money for WHA-TV as the university.

The ECB, by contrast, did not allow on-air access to the friends groups that popped up in Green Bay and elsewhere. The staff saw such organi-zations as adversaries, potential rivals for control of programming. The agency feared successful private fund-raising would jeopardize its state support. If ECB ever did seek viewer support, its staff, not volunteers, would

do the asking, and a foundation it controlled, not a private friends organization, would bank it. The ECB thought like the state agency that it was,
rather than like the community television station that WHA aspired to be.

While the WHA-TV Friends increased their fund-raising activities, their
officers continued to pursue what they saw as their first obligation, to bring
community input into WHA-TV programming. WHA program director
Dickerson decided to direct that enthusiasm into a project he called *Tryout TV*. He asked the Friends to solicit ideas from the community for television projects, to select the best project, and then fund its production. The
Carnegie Commission had defined "public" television as a grassroots institution that tapped local voices and addressed local concerns. *Tryout TV* put
that philosophy into practice, while avoiding the severe backlash unleashed
by *SIX30*. For their first project, a Friends committee chose a proposal from
a UW–Madison architectural student who wanted to attack advertising in a
documentary called *The Selling of America*. Dickerson assigned a producer
to work with the student. The resulting half-hour program in March 1975
elicited the mixed reactions expected of a nonprofessional advocating a
controversial point of view. One viewer praised the show and asked for
more like it. Another found it "a drag," boring, and lacking "pizzazz." A
third found the host smug. Being "confident about one's negativism is not
justification enough to produce such a show," he wrote.[52]

The Friends committee reviewed hundreds of proposals through the
five-year run of *Tryout TV*, suggesting a wide range of viewer interests.
Proposals included a history of magic from ancient religions to modern
entertainment; an exposé on vanity publishers from a writer who felt he
had been the victim of one; "Soap Flakes," a collage of two television soap
operas played simultaneously; an autobiographical project called "I Am a
Waitress"; and an attempt at humor titled, "Stop Crying, You Are Driving Me Crazy."[53] Not surprisingly, the productions varied in quality, but
the concept drew wide praise in the community and across the country,
including a 1975 award from the Corporation for Public Broadcasting for
community involvement.[54] Why, then, did it die after five years? Dickerson's explanation goes back to the producer-driven culture of WHA-TV.
Producers tired of producing other people's ideas, Dickerson said.[55] They
wanted to pursue their own ideas.

In the early 1980s, the ECB staff predictions about the influence of friends groups came true at WHA-TV. The most powerful person within WHA-TV was not the station manager nor the program director, nor even Ron Bornstein. Power and influence shifted to John Price, the executive director of the Friends of WHA-TV. Price had organized and directed the Community Council for Public Television for Bornstein in the late 1960s. After a decade away, Price returned to Madison in 1980 to take charge of the council's successor organization, the Friends of WHA-TV, Incorporated. He established its offices outside Vilas Hall and hired its own staff for fund-raising and promotion. Like Bornstein, Price knew that universities do not foster entrepreneurship, and he thought an independent friends organization was necessary to realize WHA-TV's national and local ambitions. About the time Price returned, however, Bornstein took two extended leaves of absence to "rescue" Washington public broadcasting organizations. Bornstein spent eighteen months as executive vice president of the troubled Corporation for Public Broadcasting and six months as interim CEO of National Public Radio as it fought to avoid bankruptcy. His absence created a void in Madison, and, with Bornstein's approval, Price filled much of it.

On-air pledge drives that began on WHA-TV in 1974 became more professional under John Price. IMAGE COURTESY OF THE UW-MADISON ARCHIVES, #S14665

Price's return to the Friends of WHA-TV marked the end of amateurism in fund-raising. He hired professional fund-raising and promotion staff and focused Friends programming support on major productions instead of grassroots access. He found an angel in Wausau Insurance to fund a national series about technology called *The New Tech Times*, taglined as "the video magazine for the electronic age."[56] In a bold break from past practice, the Friends, not the university, owned the program and contracted with WHA-TV to produce it. The arrangement sidestepped the state and university bureaucracy that often stymied innovative projects. It also put all financial risk and potential income in the Friends organization, with Price calling the shots. The producers brought in nationally prominent television journalist Mort Crim to anchor the program after the initial host, former FCC commissioner Nicholas Johnson, did not work out. Price found a second financial backer for *New Tech Times* in the Electronic Industries Association, a trade group that promoted the kind of new technologies the program explored. That potential conflict of interest raised eyebrows among some WHA-TV staff, but *New Tech Times* was a Friends project and Price argued that the trade association had no direct influence on specific program content.[57] Eighty stations across the country carried *New Tech Times* during its five-year run.[58]

Price made another bold and independent move for the Friends group in Madison. The group applied for and promised to fund a second television station in Madison, albeit with a low power signal. The Friends, rather than the UW regents, would own the station and have legal responsibility for its operation, although the organization would presumably contract with WHA-TV to provide programming. While the application never came to fruition, the support group had moved beyond mere friendship to something like control.[59]

Bornstein placed such confidence in Price that he put the Friends in charge of fund-raising and promotion for WHA radio as well. That move provided the opportunity to combine the separate radio and television program guides for Madison into a single glossy magazine supported, he hoped, by advertising. Price's vision for *Airwaves* went beyond just a program guide to a full-fledged city magazine, a project that proved far more challenging and expensive than he had anticipated. Intended as a source of income for the Friends, *Airwaves* proved to be their undoing. The magazine emptied the Friends' coffers and required a loan from the UW system to cover its losses. Price departed, and WHA-TV eventually brought the fund-raising and promotion staff back onto the Extension payroll, into Vilas Hall, and under station management.[60] Radio was allowed to go its own way in promotion and fund-raising.

Reining in the independence of the Friends signaled the passing of the Bornstein era. The primary force in creating public broadcasting in Wisconsin took his administrative prowess to vice presidencies in the UW system, ultimately running all system operations as executive vice president. His departure allowed a less confrontational relationship between WHA and the ECB, which, in turn, cleared the way for the emergence of Wisconsin Public Television three years later in 1987.

7

AND RADIO

1968–1979

"W e're getting ready to stand in line at this new Washington soup kitchen when and if it opens."[1] So wrote radio director Karl Schmidt to the WHA staff in September 1967. He had just learned that the US House of Representatives had amended its version of the Public Television Act to include radio. UW president Fred Harvey Harrington was among those who had urged Congress to insert the words "and radio" into the legislation proposed by the Carnegie Commission on Educational Television.[2] In April of 1967, he testified that the main purpose of educational radio in Wisconsin had been formal and specialized education, but "many of our program hours are devoted to the radio equivalent of what the Carnegie Commission has appropriately named 'Public Television.' I guess we'd call it 'Public Radio,' for it certainly is a service which the public seems to appreciate."[3] To prove his point, Harrington read letters from enthusiastic radio listeners throughout Wisconsin.

The Wisconsin State Radio Network did indeed attract a significant number of enthusiastic listeners, but few other educational radio stations could say the same. Radio had remained the first love of McCarty and Engel and, unlike managers of other radio and television operations, they refused to cannibalize radio to build television. Wisconsin boasted the largest budget for radio of any operation in the country and the strongest cadre of radio professionals. Radio commanded about 40 percent of Wisconsin's joint radio-television budget. Some other joint operations gave radio as little as 10 percent. Nationally, educational radio had not enjoyed the massive support the Ford Foundation had invested in television stations, such as WGBH in Boston and WNET in New York. That infusion of funds prepared those stations to produce programs for national audiences. Public radio, by contrast, was starting with small, low-budget stations and needed to establish a strong centralized national network. National Public Radio would offer free, high-quality programs for the stations' use, particularly in news and public affairs.

The decision to centralize proved key to the ultimate success of public radio as a national institution, but a few voices spoke up for a decentralized system and none spoke more eloquently than the enchanting baritone of Wisconsin's radio manager, Karl Schmidt. He, of course, saw WHA as a national production center. In the earliest discussions of a national radio system, Schmidt said public affairs programs could come from anywhere, not just Washington. As an example, he cited reenactments by WHA staff members that condensed twenty-seven hours of hearings on the dangers of nuclear radiation to nine hours.[4] WHA staff worked from edited transcripts of the hearings, re-creating the sound of a hearing room by adding reverberation and inserting coughs. Schmidt's argument for decentralization was implicitly an argument for traditional radio studio productions with actors, narrators, and sound technicians. Public radio would replace the actors with reporters, the narrators with interviewers, and would leave the studio for actual, rather than created, experiences.

While Schmidt failed to prevail on the principle of decentralization, the largest educational radio station in the country received the first radio grant from the new Corporation for Public Broadcasting. WHA would house a center for experimentation with new audio techniques, particularly binaural recording, which literally placed the listener in the middle of an audio scene, but only if heard on headphones. On normal radio speakers, voices and sounds seemed distant and jumbled. Binaural sound proved impractical for broadcast radio, but the audio experimentation center did create two lasting spinoffs. WHA's longtime music director and studio organist Don Voegeli composed commercial jingles and original music for radio programs. As the National Center for Audio Experimentation wound down, CPB provided support for Voegeli to write royalty-free original music for public radio and television nationally. The most enduring tune from this project is the *All Things Considered* theme still in use.

Music director Don Voegeli wrote the theme still used by NPR's *All Things Considered* in his electrosonic studio in Vilas Hall. IMAGE COURTESY OF THE UW-MADISON ARCHIVES, #S14739

The larger spinoff supported Karl Schmidt's lifelong interest in radio drama. Originally from Ohio, Schmidt came to Wisconsin as a student in 1941 because WHA did more and better drama than the Ohio State University radio station.[5] When the ECB hired its own executive director rather than continuing Jim Robertson's joint appointment, UW Extension consolidated its reduced radio and television operations under television's Ron Bornstein. Schmidt lost his director of radio title, but Bornstein sold the Corporation for Public Broadcasting on creating a major radio drama center in Madison, with Schmidt as its director. This development did not please the management at NPR, which believed it should receive all of CPB's national radio programming money. Stealing two hundred thousand dollars of "their" CPB support made Bornstein persona non grata at NPR, but he believed no less than Schmidt that, as in television, CPB should spread its radio funds among a few major producing stations, Wisconsin among them, of course.

Thus, *Earplay* was born. Schmidt had in mind a full range of dramatic programs like those that dominated commercial radio into the early 1950s. They included soap operas and popular drama, as well as the classics. Schmidt said he wanted to produce popular "Norman Corwin–type dramas," but pondered where he could find experienced writers. No one had written original radio dramas in almost twenty years. He finally found Peg Lynch, whose *The Couple Next Door* series had lasted on commercial radio until 1956 and who was willing to write more about the comic adventures of "Ethel and Albert." *Earplay* produced some new episodes, but, according to Schmidt, "it didn't work out."[6]

The dearth of radio writers led Schmidt and his collaborators to turn to writers in other media, particularly playwrights. An additional two hundred thousand dollars from the National Endowment for the Arts (NEA) cemented the strategy of linking radio to the stage. The link provided an inexpensive outlet for new, and often risky, plays without the cost of mounting a full theatrical production. A project conceived as reviving a popular form from the past became a place for experimenting with often-difficult contemporary drama. Under the theatrical model, *Earplay* attracted major playwrights like Arthur Kopit, Archibald MacLeish, David Mamet, and Jack Heifner. Novelists John Gardner and Donald Barthelme also wrote plays for *Earplay*. Collaborating with the BBC and other European broadcasters,

Playwright Arthur Kopit, left, works with director John Madden on *Wings, Earplay's* biggest success. IMAGE COURTESY OF THE UW-MADISON ARCHIVES, #S14673

Earplay commissioned *Listening*, a "chamber work" for radio by playwright Edward Albee. Three nameless characters—a man, a woman, and a girl—wove together strands of conversations, suggesting that communication fails even when we "listen." The stereo production emphasized that lack of communication by placing the man and the woman on separate speakers. It was an artistic success, perhaps, but not necessarily sparkling radio entertainment. The stereo recordings delivered to stations on disc before public radio enjoyed satellite interconnection drew wildly mixed reactions from stations. Many praised the artistic and technical standards. Others agreed with the audience research consultant George Bailey, who replied "1938" when asked at a meeting of program directors about the best time to schedule radio drama.

A review of the *Earplay* series in the *Los Angeles Times* captured both the strengths and the limitations of the project. Critic James Brown praised its commitment to the "present tense as a viable art form with the capacity to grow." The generally positive review spoke of the rewards of paying attention to the sounds and the voices. He was pleased it did not pay homage to the past. Pointing to what may have been *Earplay*'s fatal limitation as radio entertainment, however, Brown observed that the hour-long plays demand "our undivided attention. These plays do not always allow the comfort of familiar plots in traditional structures."[7]

Earplay triumphed with its original production of *Wings*, by Arthur Kopit. Schmidt said it worked because it combined production values with a compelling human story.[8] That production won the Prix Italia for radio drama, the first time an American production snared the highest international prize. *Wings* became a successful stage play in New York. Indeed, in January 1978, three plays that originated as *Earplay* radio dramas were on stages in New York, Chicago, and the Twin Cities. Albee advised Schmidt to move the project to New York to become an integral part of American theater. Martin Esselin, the head of drama for the BBC, countered that the theatrical approach could not reestablish radio drama in America.[9]

The conflicting advice spotlighted the dilemma that divided the *Earplay* staff. Schmidt wanted to reestablish popular drama, and he wanted to do it from Madison. His associates Tom Voegeli and John Madden, who would later gain fame as director of films such as *Shakespeare in Love*, wanted to break free of Madison and the state bureaucracy in which WHA functioned. They could not believe, for example, that the state wanted them to get competitive bids when they sought an official signature for the contract with Albee for *Listening*. The Twin Cities were not exactly New York, but they were a step closer to the theatrical big time than Madison, and Minnesota Public Radio did not present the operating hurdles of a state bureaucracy. Voegeli and Madden relocated to St. Paul, and Minnesota Public Radio took over administration of the *Earplay* project, while project director Schmidt remained in Madison.

For the next three years, Schmidt and Voegeli struggled for control of the project. The NEA helped settle the argument when it withdrew its support in 1978 after four years. Schmidt's radio entertainment philosophy would prevail in the greatly reduced project. MPR, Voegeli, and Madden were out. The project came back to Madison, but Schmidt still wanted a nonbureaucratic partner to administer the slimmed-down project. Bornstein delivered NPR, the organization that had always felt it should receive all of CPB's national programming money. Having become a forceful member of the NPR board, Bornstein softened his advocacy for a decentralized production model. He was willing to let NPR receive and allocate all national programming dollars, on the understanding, of course, that a sizable amount would go to *Earplay* in Madison.

NPR had something else to offer. Its member station at the University of Southern California had obtained the rights to a radio production of the film *Star Wars*. If Schmidt wanted to revive popular, but contemporary, radio drama, what better vehicle than the Hollywood megahit? He mulled it over for several weeks before saying no. NPR and USC wanted production to take place in Los Angeles, not Madison, and Schmidt to share control with the USC professor who obtained the rights. Besides, Schmidt anticipated a good deal of "advice" from NPR president Frank Mankiewicz, the outspoken son of Herman Mankiewicz, Orson Wells's collaborator on the historic radio production of *War of the Worlds*.

With Schmidt taking a pass, NPR turned to Voegeli and Madden to produce the series that garnered far more acclaim than ten years of *Earplay* productions. In yet another irony, NPR's overspending on *Star Wars* and its sequel, the *Empire Strikes Back*, contributed to its near bankruptcy in 1983. To regain solvency, the NPR board brought in as interim president none other than Wisconsin's Ron Bornstein. NPR had to impose massive cuts, including elimination of all performance programming, which meant the end of radio drama and the *Earplay* unit back home in Madison. Karl Schmidt retired when *Earplay* ended, but he continued to read books on *Chapter a Day* for the next thirty years.

Nationally, educational radio gained federal financial support and a new identity by trailing largely unnoticed in the wake of public television. In Wisconsin, public television dragged radio along, too, but with less happy results. The rivalry between UW Extension and the ECB was all about television but greatly affected radio, which the university had operated for many years with the passive acquiescence of the ECB's predecessor, the State Radio Council. With its own executive director and staff, the ECB would not remain passive. It had inherited from the State Radio Council all the FM stations in the state radio network, including the network flagship, WHA-FM. Only the original AM station, WHA, was licensed to the university, even though UW Extension largely funded and managed the program service for the entire network. The ECB decided in one of its earliest meetings to develop a distinct identity from UW. "The Wisconsin Educational Radio Network" replaced "the Wisconsin State Broadcasting Service" (or the State Stations) as the network's identity, and WHA-FM became WERN. The ECB employed a program coordinator for the FM network, but his options were limited because he had a small budget and had to rely heavily on whatever WHA (AM) provided from its UW Extension funding. Bornstein met the letter of the law by allocating three closet-like control rooms to feed the nine FM stations of the state network, while the daytime AM station, WHA, used the main control room and eight studios in the spacious Vilas Hall facility.

Even though television had become the medium of choice for schools and the primary interest of the ECB staff, they insisted that WERN and the other stations in the Wisconsin Educational Radio Network continue to carry about an hour a day of in-school programming. Governor Knowles's press secretary, who represented him on the ECB, was probably right when he recommended in 1969 that the governor deny ECB's request for new money for programming on the Wisconsin Educational Radio Network because he did not believe schools were actually using the programs. Even if audio still had a role in instruction, most schools recorded the programs for teachers to use when they needed them, and consequently, mailed audiocassettes provided a more efficient delivery system. While some in ECB management understood the argument against broadcast audio instruction, they decided to leave well enough alone, fearing the legislature would stop funding the FM network if it were not used for instruction. Over time, that charade ended for radio, and the argument moved to television, where new technologies provided better options than broadcasting for supporting instruction in K-12 schools.

The ECB's actions in television had another consequence for radio. The state agency built modern television transmitter systems around the state and moved the jerry-rigged transmitters of the FM network to the new TV facilities. Technically, these moves improved the FM operations, but at the price of "the weather roundup" and the "state station engineers." When radio moved from venerable Radio Hall in 1972 to the new Vilas Hall facility it would share with public television, the engineers who operated the radio transmitters and reported on weather conditions from their remote hilltops across the state disappeared and so did those three distinctive chimes that signaled station breaks.

As the ECB put "educational" in the name of the state radio network for the first time, Bornstein tried to redefine WHA as a "public" station, much as he had already done with WHA-TV. With Schmidt off trying to revive radio drama, Bornstein appointed his Michigan colleague and *Inner Core* collaborator Ralph Johnson as associate director for radio, a position he began in 1971. Together they focused WHA on serving the Madison community, particularly in news and public affairs. While WHA, WERN, and the state network carried identical programming for much of the day, WHA went its own way for certain hours to cover local issues, including programs "by, for, and about" blacks and Hispanics in the community.

This strategy proved particularly challenging for the veteran radio staff accustomed to thinking in terms of statewide educational programming and little inclined toward original journalism or issues particular to the Madison community. One well-intentioned effort tried to wed a traditional State Station technique to the newly prescribed focus of localism. WHA listeners could enjoy golden-voiced announcer Cliff Roberts reading articles and editorials from Madison-area newspapers while WERN and the rest of the FM network carried *Accent on Living*, successor to the old *Homemakers' Program*.

Whatever the reluctance of veteran staff members, some students and younger employees jumped enthusiastically into covering controversial social issues raging around them in the 1960s. Schmidt praised them for a forty-five-minute documentary they produced in one hour on the Dow Chemical demonstrations that roiled the campus in 1967. "It was quick and dirty," he wrote, "but it provided a better picture of that sad day's events than was available anywhere else."[10] WHA's younger staff were particularly proud of their live coverage of hearings that angry state legislators conducted into the violent demonstrations on the Madison campus.

It was during Ralph Johnson's tenure at WHA (AM) that my own involvement with Wisconsin public broadcasting began. I had worked with Johnson and Bornstein at the University of Michigan station, and in 1968 Johnson brought me to Wisconsin to do the kind of public affairs reporting that news director Vogelman declined to do. I left after about a year, however, to accept a CPB fellowship to work at the BBC in London and to help start National Public Radio, experiences that would bring me back to Wisconsin a decade later to lead its public broadcasting efforts as WHA radio manager. After I left, John Powell, a young television reporter from Wausau, took my slot and created a daily Madison-focused public affairs hour.

The "public" philosophy also allowed women to emerge in roles other than the homemaker's show or *School of the Air*. The very bright and driven Ronnie Hess broke the barrier by hosting public affairs programs, while Mindy Ratner added her resonant voice to the general announcing staff. But Johnson's boldest break with the past involved a new program aimed at university students, a "new" audience for public radio and its interest in diversity. Since WHA broadcast only during daylight hours, Johnson secured the cooperation of the ECB to broadcast the program on WERN late in the evening, after the other stations in the FM network had signed off. Just as television's *SIX30* provided a forum for minority and disadvantaged young journalists, *Out of Context* provided programming by, for, and about students and other young people in Madison. Joe Grant, a UW undergraduate student who had gained a following on a commercial station, created the program for WHA. He combined alternative music with a discussion of issues that interested young people. Both proved controversial with traditional educational radio listeners, including university faculty and administrators. Among those offended by the lyrics Grant played was the

A new generation of voices, including women's, found their way on the air in the early 1970s. From left to right, Mindy Ratner, Jim Fleming, and Ronnie Hess. IMAGE COURTESY OF. THE UW-MADISON ARCHIVES, #S14728

WERN transmitter engineer, who once signed the station off the air rather than break FCC rules on obscenity and indecency.

The matter came to a head when Grant complained in a November 1970 memo to Johnson about the way WHA staff members treated the volunteers who worked with him on *Out of Context*. He reported that staff members harassed the young volunteers about the length of their hair and told them to go home and take a shower.[11] The smell of pot in venerable Radio Hall in the evening may have offended them too, although Grant did not mention that in his memo.[12] Grant defended the seriousness of the topics covered by *Out of Context* in another memo to Johnson as he weighed the fate of the experiment. Grant's list of substantive "community issues"—"the status of women's lib in Madison," "nudity, go-go joints, and morality," "anti-war demonstrations," "implications of Nixon's State of the Union message for Madison politics," "Madison's drug culture," and "the occult in Madison"—may have done his cause more harm than good.[13] The day after receiving Grant's memo, Johnson wrote a "To whom it may concern" letter of lukewarm praise for Grant to take with him as the program left the air.[14]

Decisions like this took a toll on Johnson. He was more a sensitive audio artist than a manager. After several years in an uncomfortable role, Johnson stepped down. In the wings, Bornstein had Claire Kentzler, director of *School of the Air,* defunct since the ECB took over direct control of programming for schools. Her strong background in theater and instruction did not prepare her well to lead WHA into the new world of public radio as Bornstein conceived it, but she was respected and well-liked by all who worked with her. She knew that much would need to change to meet Bornstein's expectations, and the changes would prove difficult, as she later said, "because they [were] all my friends."[15]

However, Kentzler did not have much time to make the changes or damage her friendships. Within a year of her appointment, doctors diagnosed cancer, which quickly proved fatal. She was only forty-nine. Kentzler's death was the first in a series of premature passings that wiped out many of the most important figures who moved from Radio Hall to Vilas Hall in 1972. Program director Jim Collins succumbed to Lou Gehrig's disease, followed by news director Roy Vogelman to cancer, chief announcer Ken Ohst to cancer, keeper of the McCarty flame Mary Macken to cancer, and Cliff Roberts to a stroke while hosting *The Morning Concert.* Some attributed the wave of deaths to working thirty years in a building that had housed a nineteenth-century coal-powered heating plant. Others blamed the *Mad Men* cigarette and alcohol culture of that building once it became Radio Hall. More frequently, staff members spoke of the demoralization that accompanied the move from comfortable old Radio Hall to sterile Vilas Hall, from McCarty's sense of "family" to the less personal professionalism of Bornstein, and from traditional educational radio to public radio. McCarty's State Stations were gone. Wisconsin Public Radio would rise in their place.

8

WISCONSIN PUBLIC RADIO

1976–1990

I n 1976, while working in Washington as the executive producer of NPR's
flagship program, *All Things Considered,* I received an unexpected
phone call from Madison. The caller was Ron Bornstein. I had worked
with Bornstein while at the University of Michigan and again during my
one year at WHA in the late 1960s. I don't remember the pretext for his
call, but our conversation drifted to the WHA radio manager job. Claire
Kentzler's death had left the position vacant. I sensed that Bornstein was
evaluating my interest in a return to Madison, but at that point I could
hardly imagine leaving the organization I had helped create. Bornstein
was persuasive. Three months after his phone call, my family and I left
Maryland on a warm autumn day to arrive in Madison in the midst of a late
October snowstorm. As snow covered our U-Haul, I was still questioning
the wisdom of the move.

Just as when he hired Dick Lutz to shake up WHA-TV, Bornstein hired
me to revamp radio. He wanted me to turn WHA into a nationally recognized *public* radio station with an emphasis on news and information.
Using an influx of funding from the Corporation for Public Broadcasting, I
moved quickly to hire a five-person news staff and, somewhat arrogantly,
to replace hour-long programs with short segments assembled in magazine formats modeled on *All Things Considered.* My efforts provoked angry
reactions from some university leaders and longtime listeners who feared
we were rejecting the Wisconsin Idea, a controversy described in this
book's introduction. Bornstein happily ran interference with these critics.

Bornstein, however, was less comfortable with my effort to make peace with the ECB staff that controlled WERN. Early in my tenure, I realized that I had to add this effort to my agenda. Since changing the call letters of WHA-FM to WERN, the ECB had tried to establish a separate identity from WHA, while still drawing most of its programming from WHA's schedule. While WHA and WERN did broadcast different programs for parts of the day, the two stations remained essentially identical in sound and purpose. I thought each station should have a distinctive character and appeal more consistently to different audiences. I proposed that we move most talk programming to WHA, the AM station, and all music and cultural programming to the FM station, WERN. Commercial radio had used consistent formats in the United States since television became the dominant mass medium, but traditional educational broadcasters continued to mix music programs with talk programs.

My year as a CPB fellow at the BBC in London convinced me that consistent and focused formats made sense for public service radio. During that year, the BBC, the world's leading public service broadcaster, reshuffled its programming into four complementary channels, one each for rock music, light entertainment and sports, classical music, and news/talk/ spoken word. When I returned from London and went to Washington as NPR's first employee, I advocated for public radio in the United States to adopt the BBC's approach to programming. I drafted a strategic plan that called for two complementary public radio stations in each community, one built on classical music, the most popular program category among listeners to educational radio at the time, and another built on news/talk, which had the greatest potential for growth and for which NPR would need to do the heavy lifting.

In Madison, I had little trouble selling the complementary service idea to the ECB's deputy director, Paul Norton, who could never find a satisfactory answer for legislators who asked why the state operated two radio stations in Madison with essentially the same programming. A religious and political conservative who served as bishop of Madison's Mormon church, Norton chafed at examples of government waste in the form of political deal making, corruption, and bureaucratic aggrandizement. He also liked and understood radio. After listening to my pitch for distinct programming, Norton suggested we bury the separate identities of the ECB and UW Extension and use the name Wisconsin Public Radio for the coordinated effort. By putting different programming on each of the two Madison stations, we could provide more music to our music listeners and more news/information to our talk audiences and draw more listeners to both stations. The ECB-licensed stations in the rest of the state would receive a mixed schedule drawn from the two Madison stations. In the future, we hoped to add second stations throughout the state that could replicate the two services everywhere, but immediately, we could do it only in Madison. In July 1978, the ECB formally approved the plan. We implemented it on the bitterly cold morning of January 1, 1979.[1]

Squeezing more program hours out of existing resources required more efficiency, and, again, commercial radio showed the way. Prior to the change, a staff member in the music library selected records and wrote the scripts for all music programs. The staffer handed the records to an engineer and the scripts to an announcer to read on the air. The commercial

Wisconsin Public Radio used signs on Madison buses in 1979 to tell listeners that WHA and WERN offered complementary programming. WISCONSIN PUBLIC RADIO

radio–style "combo" system we instituted in 1979 gave full responsibility to a single person, who selected the music, played the records, and then talked about them, ad hoc, on the air. Combo allowed us to increase hours of classical and jazz programming and to create our three-hour *Simply Folk* program using existing personnel. Quickly overcoming his concern about operating the equipment, a young announcer, Jim Fleming, felt he was doing "real radio" for the first time. Unscripted, he felt he could relate to listeners more personally and more spontaneously, creating a more pleasant experience for the listener as well as for the host.[2] He far preferred combo to pounding on the glass between the studio and the control room to get an engineer to put down the newspaper he was reading and perform his next task. Veteran announcer Cliff Roberts had experience in commercial radio and took to combo operation gracefully, if not as happily as Fleming. The station's other great voice, Ken Ohst, continued recording his programs with an engineer. Music director Linda Clauder found herself hosting on the air and spinning her own records for the first time. Unlike educational radio, public radio welcomed female voices.

THIS VEHICLE STOPS AT ALL R.R. CROSSINGS

FM 88⁄7 Classical music & more.

SCHOOL'S OPEN
DRIVE CAREFULLY

WISCONSIN PUBLIC RADIO

Stretching existing resources worked well for WERN, but filling news and information hours on WHA was a challenge. NPR came to our rescue late in 1979 when it introduced *Morning Edition*, which reduced the pressure for state and local material during the all-important morning hours. Nonetheless, we simply did not have the staff to produce an all-day magazine format as I had initially planned. Call-in programming provided a solution. As with the music service, combo operation allowed a host to speak without a script and to operate the audio equipment to put guests and callers on the air. We could deal with one or two topics in an hour rather than the half dozen required for a produced "magazine." As Fleming found with music hosting, call-ins were real radio, a more personal way to relate to listeners. Philosophically, the format conformed to the admonition of the Carnegie Commission, which had defined *public broadcasting* as giving voice to the people. One of the report's authors had called it the "people's instrument."[3] The BBC and traditional educational broadcasting valued authoritative information, masterful performances, and credentialed experts for the benefit of mostly passive listeners. The new value that Carnegie added—and that NPR embraced in its 1969 statement of purpose—emphasized the experiences of real people, building community and understanding among diverse populations.[4] Call-in programs brought the public into public broadcasting while tapping the core expertise of academics and others with special knowledge.

Not everyone was pleased with this innovation, of course. When a UW–Madison mathematics professor complained about those "dumb" callers, I made the analogy to students asking questions in a class. He replied that he never allowed questions in his class because they were usually stupid. In fact, the producer/host of our first call-in also worried about the quality of the calls and had his call screener compile a "kook list" of folks not allowed on the air. When a student screener told a caller he was on the list, the caller took his case to the newspapers, kicking off a brief flurry of accusations and explanations and ending our list keeping.

In fact, we found no need for such a list as we expanded the call-in programming. Public radio had few "kooks" in its audience, and a knowledgeable guest and a skilled host can turn even the most eccentric question or comment into useful information. We had such a host in Tom Clark, who had worked at the old WHA as assistant to the program director before spending ten years in commercial radio. He returned to lead the news and talk programming of the station's new "public" incarnation. He hosted *Morning Edition* and then transitioned into a talk hour. With lawyer-like instincts, Clark could press guests to clarify and defend their assertions, while seeming to genuinely appreciate the insights and questions from callers. In a sense, the programming was built around his strengths and personality. His successor, Joy Cardin, remembers Clark sputtering, "Wait a minute! Wait a minute! Hold on!" when guests frustrated him. She described him as going after evasive guests like a bulldog, which caused listeners to love him or hate him. He acknowledged that his personal politics were liberal, but he was equally "adversarial against both Democrats and Republicans."[5]

The old *Homemakers' Hour* had disappeared as the faculty members who replaced Aline Hazard moved on. Just as we instituted the call-in format, the agricultural journalism department hired Margaret Andreasen, an experienced television broadcaster from the University of Illinois. She had never done call-in programming before, but we gave her a full hour for

Conversations with Margaret Andreasen and did not restrict her to home-maker topics. The broadened focus reflected the transition of the agricultural journalism department to "Life Sciences Communication" as well as the transition of WHA to public radio.

We paired Andreasen's show with an adjacent hour-long call-in program at noon. *Farm Show* host Larry Meiller took well to the call-in format and, like Andreasen, broadened this hour beyond agriculture to life

The talk staff, pictured ca. 1986. In the top row are hosts Margaret Andreasen, Tom Clark, and Larry Meiller; in the middle, producer Sarah Daniels and host Jean Feraca; and in the front row, Milwaukee host Jan Weller and producer Steve Paulson.
IMAGE COURTESY OF THE UW-MADISON ARCHIVES, #S14729

science. Many of the experts who formerly appeared in recorded features on the *Farm Show* turned up live in the studio to answer listeners' questions, which were more likely to deal with pets than farm animals and home gardens than crop rotation. Meiller recalls the very first call he received in the new format.[6] The guest was a veterinarian and the call came from an older woman in Stevens Point who asked how she could tell if her cat was a male or a female. The expert suggested she lift the cat's tail. The live and informal format revealed more fully Meiller's warm personality. He loved people and listeners loved him back. Later, humanities producer Jean Feraca added an afternoon hour that covered topics and featured guests totally different from those of Clark, Meiller, and Andreasen, but attracted a passionate following.

The call-in format uncovered several regular guests who proved sufficiently popular to spin off into separate weekend programs. Tom Clark found family doctor Zorba Paster, child psychologist Sylvia Rim, and auto expert Matt Joseph. Larry Meiller discovered animal behaviorist Patricia McConnell, whose advice on animal behavior often sounded similar to Sylvia Rim's on child behavior. The station offered weekend hours with Rim, Paster, and McConnell to stations across the country with some success. Matt Joseph's car show might have gone national except that two guys from Boston owned the auto advice niche. *Car Talk* was the most popular hour on public radio. Granted, Click and Clack, the "Tappet brothers" Tom and Ray Magliozzi, entertained as much as they informed, but, as UW president Glenn Frank had pointed out fifty years earlier, radio needed to entertain if it was to educate. That was why I approached Michael Feldman.

More than most educational radio stations, WHA had embraced entertainment throughout the McCarty era. *The Bandwagon Correspondence School* had added enough silliness to the morning to attract listeners who might never have tuned to the State Stations otherwise. Sports broadcasts brought in large audiences for entertainment rather than enlightenment, as did live broadcasts of variety shows from the Wisconsin Union. The same public radio listeners who revered *All Things Considered* loved Garrison Keillor and the Tappet Brothers. This fact reaffirmed my belief that the public radio audience responds to strong, quirky, and entertaining personalities and that Wisconsin Public Radio should find one.

I found Michael Feldman on WORT, Madison's "community" radio station. He was doing a three-hour show called *Breakfast Special* live from a diner. In a very informal manner, he interacted with whoever turned up at the restaurant, displaying the quick wit of a Groucho Marx. Others saw David Letterman's attitude in him. They both seemed to make light of the whole concept of the shows they were doing. From what I heard on WORT, Michael Feldman seemed to fit Wisconsin Public Radio, but initially Feldman did not agree with me. A true product of Madison in the 1960s, Feldman did not trust establishment institutions and he thought Wisconsin Public Radio was one. He felt at home in WORT's counterculture atmosphere and was not about to sell out. WORT's on-air fund drive, however, changed his mind. Feldman made his assigned fund-raising goal about halfway through his three-hour shift. Goal made, he stopped asking for pledges. WORT's business manager ordered him to keep going and make as much as possible. Feldman walked out. I asked him to lunch and we began to discuss a statewide radio program on WPR. We settled on a weekly one-hour, live-audience variety show at noon on Saturdays.

To launch *High Noon*, we bought a newspaper ad picturing Feldman in cowboy attire and labeled him "the fastest mike in the West." Unfortunately, that ad was the first time the executive director of the ECB, Tony Moe, had heard of the new show. For no better reason than to prove that he could, Moe declared that Feldman would never appear on an ECB station as long as he was executive director. Moe's declaration ended my hopes for a signature statewide entertainment show. Only WHA in Madison would air *High Noon*. In the end, Moe may have done us a favor. Feldman's style did not fit a tight variety show format. He cringed as the program producer checked her stopwatch and told him to move on to the next segment. After a year, Feldman proposed that he do something much looser, in the style he had developed at the diner hosting *Breakfast Special*. We moved the show earlier on Saturday morning, gave it two hours, and broadcast it from a gay-straight nightclub in downtown Madison's Hotel Washington. *AM Saturday* showcased Feldman's witty ad libs in informal conversations with audience members and entertainers who dropped in to perform. He also worked the telephones on what he called the "wash line." When Paul Norton replaced Moe as ECB executive director a couple of years later, *AM Saturday* was free to go statewide, but fate intervened again. The program director of WGN in Chicago caught the program, liked what she heard, and hired Feldman away.

Michael Feldman and the Tribune Company's radio station proved a poor match. His gig there ended in a year, and he and I were talking again. This time we plotted a national program similar to *AM Saturday* but masquerading as a quiz show. *Whad Ya Know?* mimicked Groucho Marx's classic *You Bet Your Life.* WHA announcer Jim Packard acted as straight man trying to bring a semblance of professionalism to the two hours. I sent a tape to NPR's program director. It made him laugh, and he placed *Whad Ya Know?* on NPR's schedule immediately after *Car Talk.*

While we were having fun creating new programs in Wisconsin, public broadcasting nationally faced a funding, and hence an identity, crisis. The Carnegie Commission that had recommended that the federal government support public broadcasting assumed that taxes and foundations would fund the system. The Nixon administration, however, declined to provide the level of federal support the Carnegie Commission had envi-

Michael Feldman, host of *Whad Ya Know?*, learned to milk a cow at the Wisconsin State Fair in 1985. At the right is his announcer, the late Jim Packard. WISCONSIN PUBLIC RADIO

sioned, forcing public broadcasters to look elsewhere for money. With its long history of state and university support, Wisconsin Public Radio was more fortunate than most public stations, which turned to their listeners and commercial underwriters out of desperation. Having more money would allow Wisconsin Public Radio to improve programs, but when we went "public" in 1979 and asked for listener support, our finances were fairly secure. In fact, ECB officials argued that public broadcasting in Wisconsin should not jeopardize reliable state funding by seeking the unreliable support of listeners and viewers. UW Extension's Bornstein, however, argued that public broadcasting needed an active constituency to advocate for it in state government and within the university. He thought this constituency would help ensure current funding sources while supplementing them with private money. The WHA-TV Friends had played an important role in fending off the ECB's attempt to take over WHA-TV's license. Viewers who had invested their own money in public television experienced a sense of ownership. Radio contributors, he said, would enjoy that same experience.

Bornstein had encouraged station manager Claire Kentzler to establish the 970 Radio Association to both advocate for the interests of WHA and offset the political power of the ECB. The association board consisted primarily of former radio employees and their friends and relatives. They loved WHA but had little interest in raising money. After my arrival, the station made a feeble attempt to gain listener support with an event envisioned as more promotional than fund-raising. Since the ECB did not allow fund-raising on its stations, we staged a live performance in a shopping mall on a Saturday afternoon, and it was broadcast only on WHA. Visitors could make monetary contributions at the performance, and between acts announcers in the studio asked for listeners to phone in contributions. The contributions were few and small. Of course, all odds were against the effort. We were broadcasting classical music on an AM radio station on a sunny Saturday afternoon, a time we selected because it would disturb the fewest listeners, whom the board and staff insisted would hate being asked for money.

The Wisconsin Public Radio partnership between ECB and WHA opened the possibility for a more aggressive effort. In the spirit of the partnership, the ECB's Norton reluctantly agreed to let the 970 Radio Association expand its mission to support both WHA and WERN. Renamed as the Wisconsin Public Radio Association (WPRA), it would act as a single organization working for the overall success of both Wisconsin Public Radio stations rather than the individual success of either. Norton agreed to "experiment" with a low-key, three-day, on-air drive on WHA and WERN, with all revenue going to the WPRA. Simultaneously, programming staff produced separate programming without fund-raising for the ECB state network stations outside of Madison.

Veteran WHA staff members shared the ECB's reluctance. They prided themselves on not stooping to the level of their TV colleagues, who, in their opinion, harangued viewers for money. Karl Schmidt wanted nothing to do with fund-raising, even though his son Dan, a student, organized the drive for us and even bought his father a gift membership in the WPRA. At 6:00 in the morning on the first day of the drive, I sat at the number one phone waiting to take the complaints from angry listeners. The phone rang, and the caller was not angry. He wanted to pledge. We made the entire fifteen-thousand-dollar goal for the three-day drive the first day. We tripled the goal to forty-five thousand dollars and made it easily. The experiment was a success. When we staged our next drive, the entire state network took part.

In 1981, Ronald Reagan became president with an agenda to cut government programs, including the Corporation for Public Broadcasting. Democrats in Congress saved the CPB from elimination but accepted a 25 percent reduction in its funding, which translated into a 25 percent reduction in CPB funding for WHA, WERN, and the state network. While the cuts would not kick in until two years later, we launched an "emergency" drive we called "Seven Days in May." It raised enough money to more than cover the CPB cuts. Tom Clark and Jim Fleming used passion and logic to lead the campaigns on their respective information and music stations. Each in his own words and with deep sincerity tapped the support and loyalty the State Stations had developed over the previous half century. Fleming found an ally in Robert Russell, who had come to Wisconsin to escape the cynicism of New York. Russell had the persuasive ability to make all the phones in Pledge Central ring simultaneously, earning him the nickname "Bingo Bob." After those seven days in May, there was no

About a decade after public television, Wisconsin Public Radio got serious about
raising money from audience members. The positive response kept volunteers busy.
IMAGE COURTESY OF THE UW-MADISON ARCHIVES, #S14665

turning back. For better or for worse, the State Stations had become Wisconsin Public Radio, and listener support had become a defining element.

While Wisconsin Public Radio initially provided two distinct program
services only in Madison, the resolution that the ECB had approved in 1978
expressed the intent to provide two program services statewide. Norton
and I hoped to accomplish this through agreements with radio stations

operated by University of Wisconsin institutions throughout the state. In each region, the university station would tie into either the "information service" programming from WHA or the "classical music and more" stream from WERN. The ECB station in the region would carry the other program service, providing two complementary services everywhere in the state. Norton had leverage because the legislature had given the ECB responsibility for coordinating all state-supported broadcasting in Wisconsin.[7] The university's Council of Chancellors had given UW Extension sole responsibility for broadcasting to the public, restricting campus-based stations to student training and/or serving campus audiences.[8]

In spite of this restriction, two campus stations that operated as training labs for journalism and broadcasting students applied for and received funding from the Corporation for Public Broadcasting and the right to carry NPR programming. To qualify for CPB support, the stations declared that they existed primarily to serve the public, not for training students. Norton did not want to challenge the two stations, at UW–Milwaukee and UW–La Crosse, on these contradictions, but he did see the Wisconsin Public Radio complementary service concept as a way to justify two state-supported public stations in the same region. Norton argued that extending the Madison model to Milwaukee and La Crosse would make better use of facilities the state already supported and attract private money by adopting listener-friendly formats.[9] The university and ECB stations in a region would each carry one of the core information or music services from Madison. The staffs of the university stations would "localize" the two core services by inserting local and regional material into both program streams. While the local university stations would lose program and financial autonomy, I argued in the proposal that this arrangement would provide the highest-quality service at the lowest cost.[10]

The first time we presented this proposal to UW–Milwaukee's station manager, Professor George Bailey, he rejected it. With a very sophisticated knowledge of radio, Bailey understood the value of consistent specialized formats but he did not accept the idea of statewide networks. He contended that radio stations needed to fit their individual markets and that the Milwaukee market was different from the Madison market, which was different still from Green Bay, or Wausau, or Superior-Duluth. "Your farm guy," I recall him saying, "does not belong in the Milwaukee market."

He thought nothing should limit his station's ability to serve Milwaukee, irrespective of what happened elsewhere in the state. He saw no reason why the ECB station, WHAD, which also served Milwaukee, should affect what he did on WUWM.

In making his case for local service, Bailey echoed the argument that Ron Bornstein had made about why WHA-TV should remain separate from the ECB television network. Stations are licensed to serve communities, not states, he had said. If that were true for WHA-TV, it was just as true for WUWM. Bornstein had argued, too, that the university could guard the editorial integrity of WHA-TV better than an agency directly beholden to the governor and the legislature, an argument that would apply to WUWM as well. Bailey, like Bornstein, essentially argued against the McCarty/Engel concept of a statewide network operated as a partnership between the university and a state agency, first the State Radio Council, then the Educational Communications Board.

Norton and I presented the plan to the UW Board of Regents. Unlike the ECB, which had quickly endorsed the statewide dual service plan, the UW regents gave it a "chilly reception."[11] Bornstein remained silent on the Wisconsin Public Radio proposal, but he had previously convinced the regents to protect WHA-TV from the ECB, a principle easily extended to the university's radio stations on campuses across the state. Bornstein's UW Extension boss, Dean Luke Lamb, told a Milwaukee regent that he had reservations about the proposal.[12] After our presentation, the regents accepted the university system president's recommendation to delay action on the proposal, a delay that proved indefinite.[13] Without the regents' endorsement, Norton and I set out to seek the voluntary cooperation of stations across the state.

Norton made his first priority La Crosse, home of ECB board chairman and head of La Crosse Catholic schools Father Wisniewski, who objected strongly to the duplication he heard on stations in his home community. Anxious to reach an agreement, Norton offered the UW–La Crosse station money from the WPRA without making clear exactly what was expected in return. The two sides had different understandings of what the WPR partnership meant, and the agreement broke down with some acrimony on both sides.

I was more interested in Milwaukee, where almost a third of the state's population lived, but our agreement with UW–Milwaukee worked out no better than that with UW–La Crosse. In both cases, Norton and I envisioned tightly integrated programmatic and financial partnerships under the Wisconsin Public Radio umbrella. Both Milwaukee and La Crosse sought to retain their autonomy. Milwaukee differed from La Crosse, however, in that it served a region large enough to support an independent public radio station not affiliated with Wisconsin Public Radio. Fourteen years later, in 1996, financial necessity would force La Crosse to join with WPR, but any real integration of WUWM into Wisconsin Public Radio remained an elusive goal over the decades.

A state senator from Milwaukee, Mordecai Lee, came closest to engineering an agreement.[14] He wondered why three tax-supported radio stations serving Milwaukee—the ECB's WHAD, the university's WUWM, and a school board station, WYMS—were all playing classical music at the same time each morning, and he asked the Legislative Audit Bureau (LAB) to investigate. The LAB report recommended "one entity . . . to coordinate, manage, operate, program, and promote public radio services."[15] It did not recommend consolidating licenses, but it suggested the ECB and university could operate cooperatively as if they were one entity, as they did in Madison. It endorsed dual statewide networks that would offer as much local programming as possible.[16] In other words, it endorsed the concept of a statewide Wisconsin Public Radio.

In response to the LAB's finding "considerable duplication of programming," WYMS and WHAD modified their programming, while WUWM continued its locally produced mix of NPR news and classical music.[17] WYMS went all jazz. The ECB's WHAD dropped its classical music and carried full-time the news and information format we were already producing for WHA in Madison.[18] Four years later, however, WUWM unilaterally dropped its seven hours a day of classical music and adopted an all news and information format because, its manager explained, "news and information are the main reasons that listeners make contributions to WUWM."[19] With any semblance of cooperation gone, WPR opened its own office and studio in a new downtown Milwaukee office building. It provided a home for an evening call-in talk show hosted by arts reporter Jan Weller.

Four years into the Wisconsin Public Radio partnership, we had failed to gain support from the UW system and had generated hard feelings with

ECB executive director Paul Norton, right, and radio director Jack Mitchell, center, visit WPR's new Milwaukee office, where regional manager Bill Estes showed them around.
COURTESY OF JACK MITCHELL

UW–Milwaukee and UW–La Crosse. More serious than those frayed relationships, however, was the continuing tension between the Extension and ECB staffs. The fundamental conflict was over television, of course, but it generated a deep distrust of the university among the ECB staff in all areas, including radio. Indeed, one staff member told me that the only person at the ECB who supported the Wisconsin Public Radio partnership was Paul Norton. In 1984, I realized just how close the Wisconsin Public Radio partnership was to falling apart. An ECB engineer, whom I perceived as naïve but who may have been shrewd, showed me with pride the radio studio he was building at ECB's new off-campus offices. Ostensibly for nonbroadcast uses, the studio was fitted out to broadcast standards. From that studio the ECB could provide, without any university assistance, an inexpensive statewide classical music service on its stations. Of course, without the ECB, WHA would face producing a very costly information service while having no high-power ECB transmitters to deliver it statewide. Wisconsin Public Radio could work only by combining the UW Extension production capacity with the ECB's high-powered transmitters across the state.

Anxious to save the cooperative venture we had created, I proposed to Norton that I take on the director of radio position for both the ECB and UW Extension. I was proposing an arrangement similar to McCarty's joint appointment as head of radio for both the university and the State Radio Council. I knew that Bornstein was less than comfortable with my close relationship with the ECB's Norton, but I saw no other way to guarantee the future of Wisconsin Public Radio. Bornstein, who was about to leave public broadcasting for university administration, told Norton he would not stand in the way. Norton's memo of offer to me questioned whether I could have loyalty to both organizations.[20] I said I could. Bornstein told me he approved my appointment, but I should remember that my first loyalty remained with the university. I replied that I had to divide my loyalty equally. He disagreed, he said, but respected my position. I moved my office to the ECB building, leaving Jim Fleming in charge at Vilas Hall. That radio studio I had been shown mysteriously disappeared.

Now representing both the university and the ECB, I turned my attention to parts of the state other than Milwaukee and La Crosse. University stations elsewhere lacked professional staffs and National Public Radio programming and might see integration into WPR as being in their best interest. UW–Green Bay was the easiest sell. It had built WGBW expecting to make it a public radio station for the region, but changing priorities had put that idea on hold until WPR came knocking. Using special funding from UW Extension's chancellor to strengthen relationships with UW campuses, WPR took WGBW off the budget of UW–Green Bay. We added a local host/reporter for *Morning Edition* in Green Bay. That job went to Joy Cardin, a former Green Bay television reporter. She was impressed with the level of professionalism Monika Petkus, the WPR news director in Madison, demanded. As local host, she had to arrive at the station at 4:30 in the morning, edit the state news to highlight stories of particular interest to her region, operate the equipment, give the local weather forecast, and conduct two live interviews.

WGBW had a much smaller coverage area than the ECB's Green Bay station, WPNE-FM, so we needed additional stations to cover the Fox Valley. Lawrence University had long operated a campus-based radio station, which the institution no longer wanted to support. It eliminated its station manager position and agreed to let WPR manage and program its station. We also set our sights on the UW–Oshkosh station. UW–Oshkosh, unlike Green Bay or Lawrence, had a vibrant academic program in broadcasting. It did not have enough students, however, to staff the station all day at a reasonable quality level. UW–Oshkosh could solve this problem by turning over half the broadcast day to WPR and concentrating its student talent in the other half of the day. WPR took the all-important morning hours when students were in class or in bed and let the students handle afternoons and evenings.

Several other UW campuses had FM stations tied to academic programs like the one at Oshkosh. Changing campus priorities put all of these programs in jeopardy, and WPR announced its interest in absorbing any that were willing. The UW–Stevens Point faculty made clear its intention to keep its station independent of WPR, even after it dropped its academic program. Top administrators at UW–Platteville and UW–Whitewater wanted to unload their student stations on WPR, but ran into opposition from their respective faculties. The UW–Eau Claire faculty was also protective of its student station, a viable position until the campus dropped its academic program in broadcasting, leaving the station solely in the control of students. Embarrassed by what she heard on the largely unsupervised student station, Eau Claire's vice chancellor steeled herself to student protests and turned the station over to WPR.

The handover of the student station at UW–Stout proved even more dramatic. One day in 1991, an inspector for the Federal Communications Commission paid a surprise visit to the station. The inspector found a number of violations and issued a fine, demanding immediate payment. The student on duty was in no position to pay the fine, of course, and took the inspector to the chancellor's office. The chancellor paid the fine and decided that the students really didn't need a radio station. Together, WVSS at Stout and WUEC at Eau Claire provided reasonable second service coverage in West Central Wisconsin. WRFW, the UW–River Falls student station, followed the Oshkosh example of sharing broadcast hours with WPR.

State Senator Bob Jauch brought our second service to far northern Wisconsin. He pushed a measure through the legislature to allow UW–Superior to build a high-power radio station in the Twin Ports. He also provided funding to Wisconsin Public Radio to employ the station's staff and operate it as part of our statewide dual service.

By 1990, Wisconsin Public Radio provided two services in all parts of the state except Milwaukee and La Crosse. A local journalist hosted regional editions of *Morning Edition* from offices in Milwaukee, Green Bay, Wausau, Eau Claire, and Superior. Those regional journalists also presented news to the entire network. Regional managers in each of those offices engaged their communities and helped shape Wisconsin Public Radio decisions, while development staff in those offices raised funds for the statewide service. Twelve years after Norton and I had proposed our plan in 1978, Wisconsin Public Radio provided most of the state with localized dual service, funded by a combination of tax and private money.

9

WISCONSIN PUBLIC TELEVISION

1987–2016

Not long after approving my joint appointment to the ECB and UW Extension in 1984, Bornstein took his administrative prowess to a vice presidency in the UW system. He left direction of WHA radio and television in the more academic and cautious hands of his former UW Extension supervisor, Dr. Luke Lamb. Trained in continuing education, Lamb understood that public broadcasting differed from educational broadcasting, but he was never quite comfortable with all that implied, particularly with any suggestion of commercialization. Highly responsible fiscally, he discouraged taking risks and insisted that public radio and television live within their means—and he foresaw those means declining. The state was reducing support for the university and for state agencies such as the ECB. Lamb and his ECB counterpart, Paul Norton, worried that the state would cut public broadcasting even more severely as public broadcasting drifted further from traditional education.

In 1983, President Ronald Reagan had cut federal support for public broadcasting by 25 percent and advocated ending all support, a position that congressional conservatives adopted and still champion. The Reagan administration had envisioned private supporters taking responsibility for public broadcasting as government subsidies declined. They argued that cable and satellites provided programs that were once the exclusive province of public television—programs for children, extended public affairs, science, nature, the arts, cooking, home remodeling, and other "how to" shows. Viewers paid for these cable services, the argument went. If viewers wanted what public television had to offer, they should pay for its shows as well. The public interest or the common good played no role in these calculations. Unfortunately, the call for audiences to pay a bigger part of the cost of the service they used came exactly when viewers were using less public television *because* of all those cable and satellite alternatives. Public television viewership declined, which meant fewer people contributed to on-air pledge drives. WHA-TV had nowhere to go but down from the height of its popularity in the early 1980s. The five ECB stations still had room to increase viewer support, but the ECB did not have the staff, or the moxie, to tap that potential effectively.[1]

Both men of goodwill, Lamb and Norton understood the inevitability of integrating the ECB and WHA-TV operations as budgets tightened. The threat of continuing viewership and financial declines provided the urgency Lamb and Norton needed to push for what each knew was desirable. Both administrators were there when the legislature created the current cumbersome structure and both understood the genius of McCarty's joint appointment strategy, a strategy that the legislature discarded when it created the ECB. Achieving a more functional structure would require facing the strong resistance of their staffs as well as some ECB board members' hostility toward the university.

The Wisconsin Public Radio partnership demonstrated that one person heading radio for both the university and the ECB could create a coherent statewide organization beyond the capability of either individually. Television, however, posed a bigger challenge. Norton and I were able to assemble the radio puzzle in part because neither the university nor the ECB cared much about radio. Each, however, cared a great deal about television. They resented each other's very existence, even though their similarities far outnumbered their differences. It took five years of negotiation to create a Wisconsin Public Television that paralleled Wisconsin Public Radio. A more decisive WHA-TV station manager might have achieved the agreement more quickly, but former UW–Madison professor Richard Lawson was less an administrator and negotiator than an astute critic and philosopher. Nonetheless, with prodding from Lamb, and in spite of resistance from staff, he was able to announce to them, six weeks after the fact, that Wisconsin Public Television had come into existence on January 1, 1989.[2] The ECB's Norton did not provide an explanation to his staff for another three months.[3] He said in his memo that television needed to follow the successful example of radio, with both sides "willing to define and promote WPT as a joint service of ECB and Extension."[4] This consolidation trimmed half a million dollars from television operating costs, which Norton said should finance an increase in "local programming" from 6 percent of the weekly schedule to 10 percent.[5] More important, the merger gave rise to a new identity and a more focused mission for the merged organization, Wisconsin Public Television.

The key architect of WPT was the ECB's television program director, Byron Knight. When the WHA-TV program director position became vacant, Knight added WHA-TV programming to his portfolio. He had earned the respect of WHA-TV manager Richard Lawson, who saw in him the passion of the quintessential program director. Having a single program director made sense, since WHA-TV and the state network played identical programming most of the week. Knight rarely ran into a problem serving two masters in his dual role. One notable exception occurred when PBS offered *Tongues Untied*, an autobiographical documentary by a gay black man, in 1989. Recognizing the sensitivity of the topic, Knight scheduled the film late in the evening. "The program was not especially good or important," Knight recalled, "but once it is in distribution you have to carry it or look like prudes or censors."[6] Norton, however, was willing to accept those labels, arguing that outside of Madison, Wisconsin residents would find *Tongues Untied* offensive. Knight was forced to schedule an alternative program on the state network while Madison viewed the controversial documentary.

Lawson's retirement in 1990 gave Knight the opportunity to take full control as director of television for the ECB and UW Extension. From that position he was able, finally, to provide a positive vision for Wisconsin Public Television to replace the "times are tough" rationale of Lamb and Norton as they pushed the merger. WHA-TV would no longer focus on Madison. Rather, Wisconsin Public Television would serve the entire state. Everything on the service would embody a statewide focus. "Madison lost its local station, but the state gained a stronger service," explained James Steinbach, whom Knight recruited from the Twin Cities as WPT's director of programming and production.[7] Steinbach's calm demeanor complemented Knight's more passionate style.

Longtime WHA employees resisted the changes longer and more

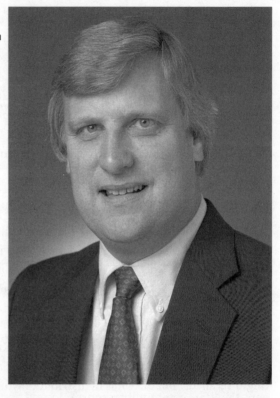

Byron Knight emerged
as architect of Wisconsin
Public Television
when the university
and the ECB agreed
to merge operations.
IMAGE COURTESY OF THE
UW-MADISON ARCHIVES,
#S15274

strongly than their ECB counterparts, but ultimately most accepted
Knight's vision for Wisconsin Public Television. Indeed, they helped re-
fine it. Knight and Steinbach convened staff meetings to discuss WPT's
role in the increasingly competitive media environment. Self-indulgent
projects would need to give way to unique and important services that public television, with limited resources, could provide the state. They began
not with what programming WPT should do. Rather, they discussed how
they wanted to be perceived by the public. The desired public perception
would, in turn, shape programming. The meetings determined that the
public should see WPT as "a state community resource" that "celebrates
and connects all the people of our state." The public should recognize
public television for its diversity and educational value. And, unlike commercial, cable, and on-demand providers, WPT should be recognized as
a free, noncommercial service that belongs to everyone. These concepts
would guide program decisions and "signal the value" of WPT to university

and state officials, and to the public, those who watched and even those who did not watch.[8]

Like public television stations everywhere, Wisconsin Public Television relied on PBS and other national distributors for most of its programming, and certainly its most popular programming, which included favorites such as Ken Burns's documentaries and *Masterpiece Theater*. National programming consumed a large part of the WPT budget and earned nearly all of its on-air pledges. Most viewers watched public television for those national offerings and contributed to sustain them. Indeed, by the 1990s, most public television stations across the country could afford to do little more than retail the national program service. In promising to "celebrate and connect all the people of the state," however, WPT set a far more challenging task. It needed to devote significant resources to programming by, for, and about Wisconsin at a time when funding from the government, the university, and viewers was stagnant and the cost of national programs was increasing.

When WPT began in 1989, Knight addressed this challenge by developing partnerships with other organizations that had similar missions. Together, they would produce Wisconsin programming that served the missions of both WPT and the partner organizations.[9] In seeking partners, the "state community resource" looked first to UW Extension, which provided the core institutional budget for WHA-TV. Bornstein, of course, had emphasized public broadcasting's independence from other Extension units. He argued successfully that public television deserved support on its own merits rather than as an instrument for UW Extension's continuing education or cooperative extension units. Knight thought he could maintain that principle by becoming an equal partner with other Extension units rather than simply providing technical support for projects they controlled. *The Wisconsin Gardner*, which premiered in 1992, provided such an opportunity. Over its twenty-two years, the program became very popular. WPT produced it by drawing on the expertise of Cooperative Extension and supporting its mission. The partnership served the interests of public television, UW Extension, and the viewing public, although it did not bring new money to the table. Shelley Ryan proved an unstoppable force in producing, hosting, and promoting the project. She then sold station management on sponsoring an annual gardening expo in Madison that has grown each year and produces significant income to support WPT programming. The success of the garden expo led to an annual quilt expo, spun off from the *Sewing with Nancy* series produced at WHA-TV's facilities.

Knight and Steinbach also saw a way to partner with educational broadcasting's traditional nemesis, commercial television. The Federal Communications Commission had mandated that commercial television provide more programming for kids. WPT wanted to produce a program for and about Wisconsin kids. In 1992, producer Kathy Bissen piloted *GET REAL!*, a program that might serve the interests of both. WPT would produce the program by and about the state's young people, and commercial stations would broadcast it. The commercial broadcasters paid only a token fee to WPT, but they delivered far more kids to view *GET REAL!* than would see it on WPT alone. The ability to reach large numbers, in turn, attracted financial support from the National Science Foundation. The foundation would not have funded the series if it could not promise a large audience.[10] It was a win on all sides. Of course, the program could not have happened without NSF's money, which was a limited commitment. Limited commitments were a problem with nearly all major projects undertaken with outside money. When the grant ended in 1998, so did the program.

As the Knight-Steinbach partnership began in 1990, they saw an opportunity to solve a national public broadcasting issue while helping WPT. The Corporation for Public Broadcasting and PBS faced criticism because they worked so closely and exclusively with a handful of major public television stations, such as WGBH in Boston and WNET in New York. Second-tier producers like Wisconsin Public Television received little national money or national airtime on PBS. Knight suggested a way for the national organizations to broaden their range of producing stations while maintaining quality. They should fund the Wisconsin Collaborative Project to produce documentaries pulling together the work of multiple stations on a specific topic.[11] The project began in 1989 with *Follow the Flag*. This program about patriotism built on contributions from public stations in Minnesota, Iowa, Chicago, and Nebraska, as well as Wisconsin. Dave Iverson, a WHA-TV producer, led the project. He remembers most proudly, however, collaborative coverage of the Great Mississippi Flood of 1993. Public television stations up and down the mighty river pooled their coverage. The program

Shelley Ryan created *The Wisconsin Gardner* in cooperation with UW–Extension.
JAMES GILL/WISCONSIN PUBLIC TELEVISION

was assembled in Madison. It gained national distribution and a strong viewer response.[12] In all, the Wisconsin Collaborative Project produced nine documentaries for national audiences before the grant money ended. Knight and Steinbach made the tough decision, then, to forgo any dreams they might have had about national productions. Instead, they would focus their resources squarely on programming for Wisconsin.

They saw statewide "public affairs" coverage as the most fundamental service Wisconsin Public Television could render. It was, they thought, the best way to "signal WPT's value" to the people and leaders of the state. Fortunately, they had on staff an exceptionally thoughtful and effective producer and anchor in Dave Iverson. Iverson left commercial television because he thought television could make a positive difference in its communities, and that Wisconsin Public Television would give him a chance to make that difference.[13] All agree that Iverson played the key role in what became the most dynamic period in television public affairs programming in Wisconsin. A strong journalist on air and off, Iverson was more philosophical than most. He thought deeply about the implications of his work for society and democracy. Steinbach acknowledged, "He drove the place."[14] Knight said, "He was the franchise."[15] They turned him loose in 1989 to create two significant public affairs projects, *Weekend* and *We the People, Wisconsin*.

Iverson wanted to do more than just produce a good television program. He wanted to make a positive difference in the state's political discourse and raise the public's engagement with issues that affected them. Iverson said those goals required taking risks, especially in an environment of declining resources.[16] He convinced Steinbach to put all available public affairs resources into one basket. The basket was *Weekend*, a live, one-hour news and interview program every Friday night right after the *NewsHour* from PBS. Iverson's young collaborator, Andy Moore, says they sought to "wake up the *NewsHour* audience" with something much more lively and a bit quirky.[17] Moore proposed positioning it as "a different kinda news show," a phrase more flip than Iverson preferred, but acceptable as long as everyone understood that the operative word was still "news." Said Moore, "If my gimmicks get more people to watch, I make no apology. I want a ton of people to watch."[18] Former WHA-TV program director Larry Dickerson was "unimpressed" with the "different kinda news show."[19] He described the result as talking heads, more radio than television. He delivered the ultimate putdown for TV purists, who believe pictures define television. Of course, the talking heads said provocative things, and the program moved along quickly from segment to segment, with constant camera movements and music transitions. The "anti-*NewsHour*" sought a sense of fun through its hour of mostly serious topics, similar, Moore said, to the whimsy and curiosity with which the early *All Things Considered* carried listeners through some heavy topics. It worked.

Iverson credits much of the program's success to the relatively civil level of political debate in Wisconsin at the time. He recalls the open exchanges between Republican Tommy Thompson and Democrat Tom Loftus, who actually enjoyed bouncing ideas off each other.[20] Other political figures joined in discussions, sometimes intense, but never angry, and always with the sense that the protagonists might head out after the program to have a drink together. *Weekend* consistently generated news for other media outlets to regurgitate and chew on over the weekend. It sat front and center in the political life in the state.

The program's tone changed slightly after ten years, when personal reasons took Iverson back to his native California. Moore was left to carry on with a new host. A widely experienced journalist, Jerry Huffman took a somewhat tougher approach to his craft than the professorial Iverson. He created a mostly healthy tension between definitions of good journalism and the role of a public broadcasting system supported in part by state government. Huffman recalled a 2002 interview with gubernatorial candidate Jim Doyle to illustrate the conflict.[21] As agreed, Doyle outlined to the *Weekend* audience his plans for education in the state. Huffman followed up by asking how Doyle would pay for these new initiatives, a question the candidate was not prepared to answer. He told Huffman the answer was on his website. Huffman responded that he had looked at the website and found nothing about how Doyle would pay for his proposals. The embarrassed candidate had no response, and the moderator let him sit in silence for a few seconds that seemed like an eternity. Huffman felt he had treated the future governor as any good journalist would, particularly his personal hero, Tim Russert of NBC.[22] Moore thought Huffman had crossed the line, and, as soon as the program ended, stormed out of the control room to say so. Huffman admits he "was a bull in a china shop" and explained that he had spent two years training journalists in the former Soviet Union to "hold the feet of officials to the fire."[23] Moore observed, "One person's good journalism is another person's gotcha."[24] More broadly, Huffman thought WPT should do "investigative reporting."[25] He wanted to uncover

Weekend filled a large studio with multiple sets and elaborate camera maneuvers to "wake up the *NewsHour* audience." JAMES GILL/WISCONSIN PUBLIC TELEVISION

what people in power preferred not to talk about. In contrast, public broadcasting from the days of McCarty had provided a forum to discuss what people in power did want to talk about.

The interviewing style of the program host mattered much less in the other major public affairs format of the era, *We the People, Wisconsin*. It put ordinary people front and center. Steinbach admits to stealing the idea from his former station in the Twin Cities, and Iverson ran with the concept in his quest to do journalism that made a difference.[26] Soon after the program began, Wisconsin became a leader in a national movement called "public journalism" or "civic journalism." This movement was an attempt to "save" journalism by making it more relevant to the real needs and interests of people and to "save" democracy by empowering citizens. The first *We the People, Wisconsin* pitted Bill Clinton against Jerry Brown in Wisconsin's 1992 Democratic primary. Citizens asked the questions. Over more than a decade, subsequent programs centered on other elections and, oftentimes more interesting, ongoing issues such as race, health care, and even land use, a seemingly technical question about which many citizens have strongly felt and well-informed opinions.

Consistent with the "partnership" theme that engulfed WPT, *We the People* brought in collaborators to increase the impact of the project and attract additional funding to pay for it. The *Wisconsin State Journal* signed on. It provided reporting on candidates and issues before the centerpiece live TV programs, and it followed up with reports about the program. Wisconsin Public Radio carried audio of the TV programs and provided reporting and discussion before and after the shows. WISC, Channel 3, added a commercial television partner in Madison, extending the reach and impact of the project. In subsequent years, other commercial television stations and newspapers around the state followed the lead of the *State Journal* and Channel 3 and became *We the People, Wisconsin* partners. Most controversial of these partners was Jim Wood, a former Democratic candidate for governor turned public relations consultant. Wood lined up financial support for the project from a range of special-interest groups. They included the Wisconsin Manufacturers and Commerce and the Wisconsin Education Association Council, the most prominent supporters of Republican and Democratic candidates respectively. Normally, public television would have found neither of the organizations acceptable to support a public affairs program, but together they neutralized the perception problem. Nonetheless, interest group funding of the showcase public affairs project and the role of a former Democratic politician in soliciting that money concerned some citizens and some within public television.

Not only did *We the People* allow citizens to ask questions, but it sometimes allowed them to indicate whether or not the responses answered their questions. Sometimes a questioner said no, producing what Moore called "astonishing moments" in live television.[27] According to Moore, the viewing audience usually identified with a questioner's frustration, even if they supported the evasive candidate. One particularly dramatic encounter saw a student break down in tears and cry out, "How dare you!" when a candidate accused her of being a "plant" for the other side.[28] Huffman recalls another woman who became outraged when a public official suggested that citizens spend more money to help the economy. She exclaimed, "You're saying the best thing I can do for my country is go shopping?"[29] The show allowed citizens to react as human beings. They were not bound by the same restrictions that the professional journalists observed.

Governor Tommy Thompson listens to a voter on *We the People, Wisconsin,* WPT's attempt to engage ordinary people in the political process. JAMES GILL/WISCONSIN PUBLIC TELEVISION

Despite wide recognition, both *Weekend* and *We the People* declined in viewership and disappeared in 2002. Andy Moore explained that Wisconsin politics had changed. Political managers began to fear rather than welcome live interactions with journalists and the public. They preferred thirty-second paid commercials that allowed them to control the message without the fear of questions or counterarguments. The officials or candidates Moore wanted became increasingly unavailable for live appearances on Friday evening's *Weekend.* This unavailability forced him to turn to second-string stand-ins.[30] In addition, shrinking budgets forced Moore to cut back on *Weekend* in subtle ways that viewers might not notice but that diminished the polish of the showcase program. As the novelty of *We the People* wore off, the other media partners invested less effort in the project, leaving most of the work to Wisconsin Public Television. The partners continued to back candidate debates, but they lost interest in topic-based debates such as on race, health care, and land use. Those trends made producing *Weekend* and *We the People* more a burden than a pleasure for Moore. He proposed ending them.[31] The co-creator of the two series, Dave Iverson, observed, "Nothing is forever. Sometimes it is better to just let things die."[32] Management, now in the person of Knight's successor, Malcolm Brett, concurred.

Prior to assuming leadership of WPT in 2000, Brett had played a key role in saving the institution from possible extinction. Brett had come up through the organization on the business side rather than the more typical path, via programming. Starting as project manager of *New Tech Times*, he moved into underwriting sales, then headed all fund-raising activities. He went on to become executive director of the Friends of WHA-TV. This organization provided financial and political support for the UW Extension side of the partnership.[33] The threat he thwarted emerged from a federal mandate that all television stations convert to more efficient digital broadcast technology by 2003. Digital broadcasting provided higher quality picture and sound. More significantly, it gave each station the ability to broadcast multiple services within the same spectrum. The problem was the estimated forty million dollars required to convert the WPT partners' six transmitters and related production and delivery systems. Knight began a serious campaign for digital conversion with demonstrations of digital's potential for multiple channels and ancillary services, particularly for instructional applications. He presented an impressive "gee whiz" show to audiences of all kinds, most notably the UW system regents. Everything he demonstrated was indeed possible through digital broadcasting; of course, each of these features was possible in any digital technology, including the Internet, which would emerge as the dominant digital delivery system in the next decade. Nonetheless, Knight made an indisputable case for digital broadcasting, particularly when he posited the end of public television in Wisconsin if WPT did not invest in digital. The funding request should

When *Weekend* ended its run in 2002, anchors Patty Loew and Jerry Huffman, left, were joined in the final show by original anchors Lynn Sprangers and Dave Iverson.
JAMES GILL/WISCONSIN PUBLIC TELEVISION

have sailed through the legislature and the State Building Commission easily; instead it got enmeshed in a debate over the structure of public broadcasting in the state and over the Reagan orthodoxy that advocated privatizing public broadcasting.

On October 27, 1999, Governor Tommy Thompson signed Wisconsin Act 9 to create the eight-member Restructuring Public Broadcasting and Funding Digital Television Transition Committee.[34] This committee was to work out the details for a private corporation to replace the ECB and the university as owner/operators of public or educational broadcasting in the state. Thompson's proposal was far from the first time state leaders had tried to restructure the state's educational broadcasting. Indeed, Larry Dickerson counted thirty such attempts, beginning with proposals to merge WHA and WLBL in the 1930s. Virtually no one looking at the two-headed structure of public broadcasting in Wisconsin thought it was the most efficient way to operate, and when the two organizations fought, it was downright dysfunctional. And fight they continued to do, even under the WPT partnership that was supposed to resolve those conflicts. While Byron Knight had come up through the ECB side of the partnership, his loyalty switched to the university as soon as he became director of television for both partners. Knight and the ECB executive director, Tom Fletemeyer, could barely talk to one another. Ultimately, they brought in a conflict resolution specialist to mediate at their meetings. Like McCarty—and me—Knight had a missionary zeal for the enterprise he led. He felt public television served a critical public function and merited preservation and advancement at all costs. Fletemeyer, in contrast to Knight, had spent his career in the state bureaucracy. He was more concerned with process and structure than with a mission to educate and transform the world. He saw his job as advancing the institutional interests of the ECB rather than public television programming. He also looked to the organization chart that showed that he was Knight's boss, and he expected recognition as such.

Governor Thompson was probably unaware of these internal divisions when he proposed his privatization plan, a plan which, it turned out, was more private in name than in reality. The new corporation would still receive state funding and the governor would appoint five board members, with legislators providing the other four. The privatized corporation would raise private funds to replace or diminish state tax support and to help pay for digital conversion. When the Friends of WHA-TV and the Wisconsin Public Radio Association objected to the board's political nature, the governor added six seats for viewers and listeners. Both the university and the ECB opposed the plan, of course, but the governor threatened: if they did not agree to cede public broadcasting to his proposed corporation, he would block state money for digital conversion and public television would go dark. The governor told Knight explicitly that digitization was contingent on privatization.[35]

That's the point at which Malcolm Brett and the Friends of WHA-TV stepped in. They could stand up to the governor as neither the university regents nor the ECB would dare. The Friends hired a lobbyist to work the legislature, and Brett himself, as an employee of the Friends, registered as a lobbyist. Since the group was made up of "friends" of Madison's WHA-TV and since the ECB opposed such groups on principle, Brett and his colleagues campaigned only for the digital conversion money and for keeping WHA-TV in the university. Their one-sided action added to Fletemeyer's frustration.

At the critical moment, President George W. Bush appointed Governor

Tommy Thompson secretary of health and social services. He took the governor out of the fray and ended the fracas. Members of the State Building Commission nearly always followed the governor's lead. The governor chairs the commission and when Thompson was governor, he never lost a vote. Never, that is, until he became a lame duck. The vote to provide fourteen million dollars for digital transition took place at Thompson's last commission meeting in January 2001—the week he left for Washington. Thompson voted against providing the money, but the rest of the body that he once dominated did not vote with him. Public television would convert to digital, which all agreed was good, without having to restructure, which may or may not have been good. In any case, Acting Governor Scott McCallum did not pursue reorganization, nor did subsequent governors Jim Doyle and Scott Walker.

The crisis averted, WPT hunkered down into a period of consolidation and relative quiet. Brett followed Knight as director of WPT. When Knight retired in 2006 Brett became director of all broadcasting for UW Extension. Politically astute and diplomatic in style, Brett made peace with the ECB and state government while strengthening the range of partnerships Knight had begun. State Senator Gary George's corruption conviction removed the anti-university senator from his leadership role as chairman of the ECB. His removal smoothed the way for a less confrontational relationship among the partners. The appointment of Wisconsin Public Radio staff member Gene Purcell as the ECB executive director gave Brett a constructive partner on the other side. The two managed to divide responsibilities between the ECB and UW Extension. Their shared understanding of public broadcasting provided a supportive environment for the directors of both Wisconsin Public Television and Wisconsin Public Radio. Purcell and Brett even agreed to turn the Madison-centered Friends of WHA-TV into a statewide Friends of Wisconsin Public Television, an event thirty years overdue.

With digital funding approved in 2001, WPT faced the task of making good on the promises of digital television. PBS provided programming to fill one additional channel with "PBS Kids" programming and an array of "how to" offerings. These shows covered cooking, gardening, remodeling, and other skills. Companies that produced products or services related directly to the program topics sponsored some of them. While public television had pioneered this kind of programming with *The French Chef* and *This Old House*, the array of "how to" shows on this additional PBS channel differed little from similar programming on numerous cable channels. Even acclaimed PBS children's programs faced credible competition from Nickelodeon and the Disney Channel, and public television's defining children's show, *Sesame Street*, entered into a new production relationship with HBO in 2015 that put all productions on the premium channel before public television can broadcast them.

Signaling its intent to offer something truly unique to Wisconsin viewers, WPT labeled its next service "The Wisconsin Channel." Without new money to program the service, however, WPT had to limit it to free or low-cost programming. *University Place*, for example, played recordings of lectures and other events at the university and created an online archive for this material. This offering was educational television at its most basic. It copied the lecture programming that had characterized educational radio decades earlier. It did, however, cement WPT's connection to the university and provided a stream of low-cost, low-production-value programming of interest to specialized audiences. Another low-cost program, *Director's Cut*, gave Wisconsin filmmakers a chance to talk about and show their films. Most of the schedule on the Wisconsin Channel, however, consisted of repeats from the main broadcast channel.

WPT gave veteran public affairs producer Andy Moore time on the Wisconsin Channel to present local and state musicians he enjoyed. His real job, however, was producing *Here and Now*, the half hour weekly public affairs show that replaced *Weekend* in 2003. It was a slimmed-down version of the old program and aired on the main channel. Brett said the half hour hosted by Frederica Freyberg remained WPT's top priority, but the organization simply did not have enough government and viewer money to support a *Weekend*-scale project. On the plus side, WPT would not seek outside support for public affairs programs as Jim Wood had secured for *We the People*. No outside money equaled no perception of outside influence. No outside money also meant a relatively low-budget program with no aspirations to wake up *NewsHour* viewers.[36]

The inability of WPT to produce as much state programming as it would have liked for the Wisconsin Channel and the main channel reflected the major financial challenges of public television in the twenty-first century.

Costs kept rising and tax and viewer support did not. WPT allocated what funds it had to general operations, the purchase of national programming from PBS or elsewhere, and its core commitment to state public affairs programs. Anything beyond those basic services required the help of partners and project-specific funders interested in seeing particular projects advance. *Here and Now* stands alone among WPT's major offerings in eschewing outside funding. Cultural programs and the human interest features of *Wisconsin Life* all require the participation of funders willing to underwrite their costs. In what Brett called an "eat what you kill" approach, projects that attracted funding went ahead; those that did not, didn't.[37] Projects needed to fit WPT's program mission, of course, but they also needed to fit the priorities of partner organizations and others willing to pay for them.

Wisconsin Hometown Stories, which premiered in 2002, provided a classic model, and the Wisconsin Historical Society, a perfect partner. The series told the state's story through the individual stories of carefully selected communities. Society staff identified twenty-five communities that represented different strands of Wisconsin's economic, ethnic, and cultural history.[38] WPT experimented with a half-hour portrait of Portage, the crossroads of Wisconsin, where a canal connects the Wisconsin and Fox Rivers. A year later the project blossomed into a one-hour profile of the industrial city of Janesville. The program received funding from the owner of that city's newspaper and two radio stations, its local power company, its largest construction company, and several foundations with an interest in Janesville.[39] The Janesville project set the pattern WPT followed to pull together partners and funders in the remaining cities on its list of twenty-five. They explored Green Bay, La Crosse, Wausau, Manitowoc/Two Rivers, Juneau County, and Oshkosh and continued to seek money to fill out the series. In each location, WPT and Society staff brought together members of the community to uncover stories and materials for the program. In Juneau County, for example, residents expressed particular interest in ethnic stories, religion, and biographies. Steinbach judged the hometown stories project a major success because it involved many individuals in each community and attracted wide attention from the public and significant financial support from local funders.[40] ECB's staff extended the value of the project to education by preparing supplemental materials for fourth graders studying Wisconsin history. WPT made programs available for sale to the public as a permanent record of the communities.

WPT found another highly fruitful partner in the staff of the Wisconsin Veterans Museum. They expressed an urgent need to record the stories of Wisconsin World War II veterans before they were gone. Working with that museum, the Wisconsin Historical Society, and an array of veterans' organizations, producer Mik Derks and other WPT staff interviewed survivors of "the Greatest Generation." From these raw materials, they created four powerful documentaries, the first an overview and one each about the war in Europe, in the Pacific, and on the home front.[41] Each participant received a copy of his or her interview and all interviews were added to the Veterans Museum collections. Under the "eat what you kill" philosophy, this project required outside funding, which came from the Ho-Chunk Nation, a labor union, and a construction company, along with numerous individuals and foundations. The success of the World War II project led to two programs on firsthand memories of the Korean War. The Society supplemented this venture with a companion book.

These war histories made a Vietnam series inevitable. More recent and far more controversial than the "good war" or Korea, Vietnam stirred emotions like no other. Few who worked at Wisconsin Public Television had served in Vietnam, and many of those who were around at the time probably opposed it. That did not matter, however, because the project focused on the men and women who, willingly or unwillingly, actually fought there. Veterans told their personal stories in their own words. The interviews with Vietnam vets provided the meat for three major documentaries and two smaller programs for broadcast in Wisconsin plus a one-hour national version that PBS distributed. Once again, the Wisconsin Historical Society published a companion book and the veterans' museum preserved all the interviews.

But the project did not end there. Those interviewed often referenced the way they came home, not to parades and brass bands, but to silence and sometimes hostility. Some appreciated the lack of fanfare, but others felt deprived. Jon Miskowski, the chief fund-raiser for Wisconsin Public Television, took up the cause. Television programs were important, he argued, but they were not enough. To make a true difference, Wisconsin Public Television should stage a massive "welcome home" event, albeit thirty-five years late. Steinbach said the idea emerged slowly from many extensive conversations with veterans about what mattered to them most. "We could never have pulled this off without collaboration and input from veterans from the get-go," he recalled.[42] Staff did not warm to the idea immediately. Some argued that staging such an event made a statement about the veterans' concerns beyond the straightforward telling of their stories in television documentaries. Besides, it would take a tremendous amount of time and work to pull off something that was not primarily a television program. Even many of the veterans did not want the bother.[43] Miskowski persisted but might not have prevailed had he not come up with an irresistible location—Lambeau Field. WPT staff and veterans groups worked three years organizing the event, as producers continued work on the television documentaries. All came together the weekend of May 21, 2010, when seventy thousand veterans and their families converged on the Green Bay Packers' home field. On the field itself sat 1,244 empty chairs to represent the Wisconsin veterans killed or missing in action in Vietnam.

The outpouring of participants was matched by an outpouring of financial support for the Lambeau event and the documentaries. Miskowski lined up two individuals and a major bank as lead sponsors, with additional funds from three Indian nations, a defense contractor, automobile dealers, a fast-food chain, a convenience store chain, and many foundations, most notably the Lynde and Harry Bradley Foundation, best known for funding conservative causes.[44] The project could not have happened without this money, but Steinbach insists that money was never the point. "We did LZ Lambeau because it was the right thing for Wisconsin. LZ Lambeau raised a lot of good feeling about WPT and raised awareness about our value in the community." Steinbach added, "That, in turn, brought us into contact with folks who may not have thought of us as relevant to their values, needs, and lives."[45] Unquestionably, splintering audiences among hundreds of digital alternatives cut public television viewership, but making a difference in the state might provide reasons for residents to support the service even if they watch less than they once did.

Wisconsin's pioneering work in "community engagement" caused the Corporation for Public Broadcasting to establish a center in UW Extension to promote the idea nationwide. During its ten-year operation, the National Center for Media Engagement encouraged public radio and tele-

LZ Lambeau "welcomed home" Vietnam War veterans more than 35 years after the US pulled out. The empty white chairs waited for those who did not make it home.
JAMES GILL/WISCONSIN PUBLIC TELEVISION

vision stations "to discover, understand, and address community needs and aspirations . . . to build sustainable community relationships, and stimulate citizen participation."[46] The center's philosophy has become the operating philosophy of Wisconsin Public Television and, according to the associate director of WPT, Kathy Bissen, will provide the organization with direction well into the future.[47]

No project exemplified that philosophy more clearly than the 2015 production, *Vel Phillips: Dream Big Dreams*. Jon Miskowski raised money mostly from not-for-profit organizations for a project that WPT felt would

make a difference in Wisconsin. The television program told the story
of Vel Phillips, the first African American woman to graduate from the
University of Wisconsin–Madison law school, the first African American
and first female Common Council member in Milwaukee, and the first
African American in the US elected to an executive office in state government when she became Wisconsin Secretary of State. *Vel Phillips: Dream
Big Dreams* presented as history the events that WHA's *Inner Core* project
had covered as they happened in 1967. The 2015 documentary reached
70,000 viewers on its first showing, enough in itself to qualify as a success.
Community engagement demanded more, however. Working with 15 organizations statewide, WPT organized town hall meetings and community
forums that attracted five thousand participants to discuss issues raised in
the television program.[48] In the spirit of the Wisconsin Idea, Wisconsin
Public Television had moved beyond producing a television program to
fostering discussion and, perhaps, positive change.

To shoot *Wisconsin from the Air,* the production crew filmed footage from a helicopter.
Pictured here are director of photography Travis McMunn, left, and pilot Nick
McMahon, filming in Dane County. LAURIE GORMAN/WISCONSIN PUBLIC TELEVISION

The previous year, development director Miskowski assembled another group of funders to support a project that may have had less community involvement but more impact as a signature television program defining WPT's connection with the state. Its artistic and financial success confirmed the wisdom of the state focus and set up Miskowski to become the next director of Wisconsin Public Television in 2015. For a decade, staff had discussed producing a helicopter's-eye view of the state, but the idea bogged down in questions of ownership and funding. With rights questions resolved and funding in hand from several individuals, foundations, and businesses, WPT hired a helicopter crew to film *Wisconsin from the Air* for a week in October 2013. Edited and enhanced with original music, the program received its first broadcast in December 2014. Former WPT Director James Steinbach described it as "a love letter to Wisconsin," and the state returned that love. In unprecedented numbers, viewers pledged money to WPT to receive their own copies of this stunning visual experience. With LZ Lambeau, *Vel Phillips*, and *Wisconsin from the Air*, Wisconsin Public Television "signaled its value" as a service that "celebrates and connects all the people of our state."[49]

10

THE IDEAS NETWORK

1990–2016

A pudgy red-haired man made several visits to Madison in the early 1990s. He told me he was assembling program packages for something called "XM Radio," a satellite service that would someday broadcast directly to automobiles. The red-haired man said he needed to fill thirteen channels on the satellite and wanted a public radio channel as one of them. Naturally, he had approached NPR first, but public radio politics prevented the national network from competing with its member stations, so he turned to Wisconsin Public Radio, which produced more hours of programs than other public radio stations. Most stations relied on national programs they did not produce nor own. He thought WPR might create a national program service with the programs we already produced and owned. Even though most WPR-produced programming was state-focused, he saw us as his most natural partner in providing a satellite public radio channel without NPR at a relatively low cost. His plan never materialized. A restructuring at XM Radio greatly expanded its ambitions to one hundred digital channels and ended our discussions.

Those conversations with the red-haired man caused me to realize, however, that satellite services—and later the Internet—would give listeners access to unique programming without the filter of a local station. Content would have no borders. In the future, Wisconsin listeners would not need WPR to hear national or international programs. Our service would have value only to the extent that we offered Wisconsin listeners programming not available elsewhere. Similarly, unique programs produced in Wisconsin might find listeners around the world. This realization shaped all subsequent decisions at Wisconsin Public Radio.

Beginning in the 1930s, Wisconsin's State Stations had offered programs to national audiences. The practice began with *School of the Air* programs and reached full flower with the *Earplay* national drama center. As Wisconsin Public Radio, we offered national audiences programs built around unique personalities who happened to live in Madison— Michael Feldman, Zorba Paster, and Patricia McConnell. To compete in the upcoming international audio marketplace, however, I wanted to add a signature program not built on a specific personality but constructed from our heritage as the premier university-based broadcaster in the country. I sought a structure that would expand and endure even as personnel came and went. WPR should produce the most intellectual program on public radio, I thought, a program that might not draw the largest audience but that would firmly connect public radio stations with the universities that owned so many of them and would appeal to a niche audience worldwide. I described it as journalism about the academic world. It would appeal to people who wanted to explore topics that captured their interest in their college years. We christened it *To the Best of Our Knowledge.*

I staffed *TTBOOK* with our best and brightest. I reassigned news director Ann Strainchamps and senior talk producer Steve Paulson to the show. Program director Jim Fleming gave up his administrative duties to host and lend his distinctive voice and sensibility to this signature national offering. Ultimately, the trio abandoned my notion of journalism about the academic world in favor of single-topic hours that pulled together several related interviews. They retained the goal of producing the most intellectual program on public radio. Carriage by public radio stations never grew to anything approaching the level of *Whad Ya Know?*, but the program's presence on the web grew steadily, particularly as the producers focused more of their energy on their digital strategy. And, yes, it was carried on SiriusXM, a far more complex and costly satellite service than the red-haired man had described to us in the early 1990s.

It had not been our national programming that drew the red-haired man to Wisconsin, however. It was our unique Ideas Network that intrigued him. By 1989, WPR had accumulated enough additional transmitters and regional offices to finally deliver two distinct program services in

most of the state, but radio's environment had changed since 1979 when
we envisioned dual service. In 1979, people listened to AM radio and FM
radio almost equally. By 1989, FM had left AM in the dust. The only AM
loyalists were older listeners and fans of right-wing talk shows. In addition, news and information had replaced classical music as public radio's
primary appeal across the country. Our original 1979 concept was to run
the now dominant public radio format—news and information—on the
withering AM dial in Madison, and the declining public radio format—
classical music—on the newly dominant FM radio band. By 1989, I was
not comfortable with that concept. The dominant format belonged on
the dominant band.

George Bailey, former WUWM manager turned public radio research
consultant, emphatically confirmed my discomfort. Using the statistical
technique of factor analysis, Bailey reported that Wisconsin Public Radio
had three distinct audiences in Madison: a group that gravitated toward
the NPR newsmagazines, a group most attracted to classical music, and a
third group that preferred what Bailey described disparagingly as "Vilas
Talk," that is, the call-in programming and other Madison-produced programming such as the old stand-by, *Chapter a Day.* Based on his experience
around the country, Bailey suggested we put the NPR newsmagazines on
FM where listeners expected to find them. I agreed. It made no sense to put
classical music on AM radio. It demanded FM quality. Therefore, WERN
in Madison would need to carry both NPR news and classical music. As
for "Vilas Talk," Bailey advocated consigning it to WHA (AM) and allowing
them to wither and die together. The Ideas Network on WHA emerged
from this reality to complement the NPR News and Classical Network
on WERN.

Bailey and I agreed on that configuration, but for different reasons.
His research across the nation convinced him that public radio stations
did best when they "retailed" national programming, primarily from
NPR, and classical music from other continents and other centuries, all
in the name of quality. I, in contrast, revered the Wisconsin Idea and the
Carnegie Commission vision of public media as an instrument of democracy. He saw live interview/talk programming as lacking "quality"; I saw
it as the essence of community involvement. He had little use for state and
local content; I saw it as our future. Even before I met with the red-haired

man, I thought pubic radio's future would belong to those who produced unique content rather than to those who merely distributed content others produced. Finally, I was convinced that, more than most other public media, Wisconsin Public Radio had the resources to produce a large percentage of what we broadcast.

While digital technologies opened opportunities for Wisconsin to deliver programming to a worldwide audience, those same technologies opened Wisconsin to public radio programming from outside sources. I was sure the NPR News and Classical Network would thrive in the near term, but it might decline in the long term as listeners found new ways to receive its namesake programming. I feared that the unique Ideas Network would struggle in the near term, but believed it would retain an invulnerable niche in Wisconsin over time no matter what happened in the media environment.

Tom Clark thought I was too pessimistic about the short-term prospects for the Ideas Network. The veteran of the old State Radio Network said educational radio in Wisconsin had listeners for fifty years before NPR existed and would have listeners without the NPR newsmagazines. Clark brought that spirit to the three-hour morning talk show he hosted on the Ideas Network in competition with *Morning Edition* on the NPR News and Classical Network. Our other pre-NPR veteran, Larry Meiller, brought the same optimistic spirit to his expanded midday program. Between Clark's courtroom and Meiller's kitchen table, program director Jim Fleming placed the more ethereal Jean Feraca, WPR's "humanities" reporter. Her two hours provided something completely different from the down-to-earth reality of Clark and Meiller. She explored spirituality, poetry, Italian cooking, and other topics close to her heart and brought a charming and insightful, "out of the mouth of babes" innocence to more mainstream topics she cared nothing about. Her supervisor, Joy Cardin, still shakes her head recalling Jean asking, "Who's Bill Proxmire?" or "Who's Brett Favre?"[1] Her enchanting voice guided listeners through unexpected twists and turns, and thousands enjoyed the journey.

Three years later, in 1993, Kathleen Dunn joined the Ideas Network. She had been doing a public radio–style talk show on Milwaukee's top-rated commercial station, WTMJ. When a rival station placed Rush Limbaugh opposite her program, WTMJ management told her to become a strident liberal on the air. Although her personal political views were indeed liberal, she could not make herself belittle conservatives as her bosses wanted. Ultimately, she believed in reason, compassion, and fair play, attributes no longer welcome in commercial radio. She found a more congenial home at our Milwaukee studio. She was comfortable hosting a three-hour program on the Ideas Network where she could share the many resources of Wisconsin's largest metropolitan area with the rest of the state. While not ideological enough for WTMJ, Dunn did allow her personal values to show more than other WPR hosts, which drew criticism from some conservatives. Dunn did not tell her listeners what to think, Cardin said, but she did provide a warmer welcome to guests with whom she sympathized than to those with whom she did not.[2] Nor did she try to disguise her anger at injustice.

Greg Schnirring, who joined WPR in 1990 as program director, regarded the creation of the Ideas Network as the most profound change in Wisconsin Public Radio's history, but he thought the reconfiguration of the NPR News and Classical Network caused the most controversy.[3] The change meant that Madison-area listeners would wake up to hear *Morning Edition* instead of classical music. Based on the experience of other stations that had replaced classical music with *Morning Edition*, I expected about one thousand complaints. Schnirring recalls hearing from about three thousand listeners objecting to the loss of those two hours of classical music.[4] Madison media gave the protests prominent play. A highly successful pledge drive three months after the change muted the outcry. The next rating book showed WERN, our FM station carrying NPR news and classical music, with the largest audience ever, which I had expected. WHA, our AM Ideas Network station, had no drop in listeners, which I had feared.[5] Thirteen years later, in 2003, the Ideas Network surpassed the NPR News and Classical Network in statewide listenership and continues to attract 60 percent of WPR's total statewide audience.[6]

When extending the dual service concept outside Madison, we placed the NPR News and Classical Network on our most powerful transmitters in areas that did not have other sources for public radio's signature programming. We put the Ideas Network on the smaller stations in those regions. In parts of the state that had other sources for NPR's newsmagazines and classical music, we put the Ideas Network on the larger stations and NPR

Kathleen Dunn's Ideas Network talk show broadcast from WPR's Milwaukee studios in a downtown office tower. JAMES GILL/WISCONSIN PUBLIC RADIO

News and Classical Network on the smaller stations. Milwaukee's WUWM, for example, carried NPR's newsmagazines, and the commercial station WFMR played classical music full-time. In Milwaukee, therefore, we placed the Ideas Network on WHAD in Delafield, the single most powerful transmitter in our system. Similarly, we placed the Ideas Network on the larger transmitters in the western part of the state, where Minnesota Public Radio beamed in NPR's newsmagazines and classical music.

For the most part, we received little negative reaction to these changes outside of Madison, but there was one that made a lasting impression. My wife and I were having dinner in a small dining room at the Old Rittenhouse Inn in Bayfield. A group of couples entered and the host seated them at a nearby table in the otherwise empty room. One person in the group remarked that the place was lovely and asked how the host couple had found it. The wife replied that she had heard about it on Wisconsin Public Radio. (I, director of Wisconsin Public Radio, smiled to myself.) Someone else in the group jumped in, "Wisconsin Public Radio? I *hate* Wisconsin Public Radio." (I, director of Wisconsin Public Radio, ceased enjoying my meal.) "They took classical music off my station. So I wrote the manager and said if they don't put it back, I am going to listen to Minnesota Public Radio." She continued, "And he answered that I *should* listen to Minnesota Public Radio if I want classical music. Can you believe that?" (The director of Wisconsin Public Radio wanted to leave as quickly as possible.) While my suggestion that she listen to the station that gave her what she wanted made sense rationally, it did not satisfy her emotionally. She felt a trusted friend, whom she assumed shared her tastes and values, had hurt and betrayed her. This sense of trust is common among public radio listeners, and it made changes difficult. At the same time, it was the attitude that bonded public radio to its listeners.

Two decisions in the ensuing years tested that bond with two groups of listeners. In 1993, we announced that we would no longer carry live Saturday afternoon broadcasts of the Metropolitan Opera. We would continue to broadcast opera, but not the live matinée on Saturday afternoons. We had rational reasons for making this change, but the opera-loving community, particularly in Madison, was well organized and passionate. While public radio program directors across the country urged us to hold firm on the decision so they could follow our lead, opera lovers pulled out all the stops to change our minds or, failing that, to get the university or the ECB to overrule our decision. Program director, and later director of radio, Greg Schnirring recalls that Ron Bornstein had retired from the university administration and could no longer protect WPR from outside interference in the way we had come to expect.[7] Coincidentally, both the ECB and UW Extension changed leadership during this period. The new chancellor of UW Extension, Don Hanna, and the new executive director of the ECB,

Glenn Davison, arrived at their posts to find piles of communications about the opera. Opera was not particularly important to either of them, and each wanted to tend to other matters. They asked me to change the decision. I did, despite Schnirring's warning that giving in to organized pressure on this issue would encourage others to try the same tactics in the future.

Schnirring's prediction proved correct three years later when our programming team decided to pick up *Car Talk*. It was the most popular public radio program in the country, and the team planned to run it during its live national feed at 9:00 Saturday morning. That timing bumped our own locally produced car advice program. Ideas Network program director Joy Cardin gave host Matt Joseph several options. He rejected all of them. He warned her, "You move *About Cars* and you will be sorry."[8] The Ideas Network programming team concluded they had no choice but to drop the program. Any chance to return the program to our schedule evaporated when Joseph took his case to the state legislature. He went from office to office with a presentation about the success of his program. One state senator demanded that the Legislative Audit Bureau investigate WPR's decision-making process. The auditors reported back that WPR had followed proper procedures in making the decision. That report should have ended the matter, but Joseph asked for a public hearing on the report. He got the hearing, at which he and his fans vociferously aired their grievances. Both the university administration and the ECB took a passive approach and never raised a public objection to this legislative intrusion. While their failure to defend WPR disappointed me, the licensees never hinted that we should back down from our decision. Nor, despite some grandstanding by individual legislators, did the legislative committee take any action on the issue.

Car Talk was here to stay and did prove popular, but the opera and car show incidents tainted Wisconsin Public Radio's reputation nationally. When I decided to move to the UW–Madison journalism school faculty in 2008 after twenty-one years as head of WPR, Wisconsin could not attract an established nationally known figure to replace me. The ECB and UW Extension hired the business manager of a smaller public radio operation in Seattle, Dana Rehm. She brought a much-needed outside perspective to the organization. While Rehm was unprepared for the complex politics of public broadcasting in Wisconsin, she made a permanent impact on the organization by raising WPR's fund-raising to the professional standards common elsewhere in the country. During my years as director, WPR had developed a large membership base, but individual donations were relatively modest. Rehm worked to boost the size of donations up to national standards. More significantly, she began to expand and professionalize those short advertising-like announcements that public broadcasting calls "underwriting." Rehm understood how much WPR lagged behind national averages and took the steps necessary to make underwriting a significant income source.

Like Byron Knight in television, however, her activism and independence clashed with ECB executive director Tom Fletemeyer. After three uncomfortable years, Rehm escaped to a vice presidency at NPR. Extension and ECB promoted associate director Greg Schnirring to replace her. He led a careful don't-rock-the-boat regime, and he worked hard to maintain constructive relationships with the university and with the ECB. Schnirring's most lasting contributions were in the technical area. He finessed both licensees to put together financial packages to replace aging and neglected hardware and to add new stations in the few places that lacked one or both networks. He also worked with the ECB and the UW Extension structures to round up funds to add digital capacity to WPR transmitters across the state and digital production equipment to studios. His efforts set the groundwork for each WPR transmitter to broadcast three digital channels, the Ideas Network, the NPR News and Classical Network, and an all-classical network. A few years later, Schnirring followed Rehm's path to Washington with a vice presidency at the Corporation for Public Broadcasting.

Next to the director's desk came Phil Corriveau in 2004, who as a young man had worked for both the ECB as an engineer and WHA as a programmer. He had moved on to build, virtually from scratch, a major public radio operation in Sacramento, California. A stint as manager at the University of Texas radio station in Austin preceded his return to Madison and his dream job as director of Wisconsin Public Radio. Corriveau built on Rehm's innovations in fund-raising, particularly planned and major gifts. He continued Schnirring's efforts to add stations and upgrade facilities. Health issues culminating in a near-fatal stroke limited, and eventually ended, his ability to implement all his plans.

Corriveau's deputy, Mike Crane, succeeded him in 2010. Having worked in public radio stations around the country, Crane understood that the strengths of Wisconsin's peculiar structure greatly outweighed its limitations. He also knew it demanded careful handling. More gregarious than any radio manager since H. B. McCarty, Crane thrived on the emerging role of a public radio manager as the public face of the organization—to explain, to advocate, and to seek funds. He was well suited to advance Corriveau's emphasis on major gifts and planned giving. These were relatively untapped resources, particularly considering the nearly one hundred years of goodwill accumulated by WPR and its predecessors. He observed that public television was far ahead of radio in this area, which had helped the visual medium survive in the face of declines in other funding sources.[9] Crane also picked up where Rehm had left off in developing the potential for underwriting support for WPR. By 2014, underwriting produced more than $2 million a year in income for WPR, still less than listener contributions, but an indispensable part of the organization's financial mix.[10]

While his background and first love was programming—he had been on the founding board of the Public Radio Program Directors Association (PRPD)—Mike Crane ceded the top programming role to another leader of the PRPD and another "Mike," Mike Arnold, whom he lured away from the top programming job at Public Radio International in 2012. Prior to the two Mikes' arrival, programming at WPR had changed little since the creation of the Ideas Network and NPR News and Classical Network model in 1990. During those twenty years, management dealt with unavoidable changes, most notably the retirement of Tom Clark, whose personality had dominated the Ideas Network. That network's director, Joy Cardin, took over Clark's three morning hours and ultimately relinquished her management responsibilities. While she never tried to equal Clark as an incisive interviewer on public policy issues, Cardin brought to the program Kathleen Dunn's empathy and Larry Meiller's friendliness. She also chose more topics of practical concern to listeners.[11] In 2009 she added a weekly hour with financial planner Kevin McKinley that proved very popular and demonstrated the importance of WPR's regional offices. McKinley started doing a local program in the Eau Claire region and continues to participate in Cardin's program from WPR's Eau Claire studio.

When Tom Clark retired, Joy Cardin brought a less aggressive style to mornings on the Ideas Network. JAMES GILL/ WISCONSIN PUBLIC RADIO

Even though they were considered outsiders, the two Mikes decided to modernize, but not change, the essence of Wisconsin Public Radio. In particular, they decided to retain the character of the Ideas Network, still an odd duck among national public radio strategies. Arnold wanted to appeal more to younger listeners but do it in the spirit of the Ideas Network. To achieve that appeal, he designed the three-hour daily program *Central Time*. Like all Ideas Network programs, *Central Time* needed to air live and take listener calls. It differed from other Ideas Network programs by covering several topics in an hour rather than only one. To host the program, he plucked two younger staff members, Veronica Rueckert and Rob Ferrett, from other programs. Rueckert had a mellow public radio sound; Ferrett sounded edgier. Arnold rejected the generally accepted dictum that successful public radio must be expensive, tightly edited, and highly produced to attract and keep listeners. He understood that public radio's

Talk director Sheryl Gasser led the planning in 2013 for *Central Time*. Tossing around ideas were producer K. P. Whaley and program hosts Veronica Rueckert and Rob Ferrett.
JEFFREY POTTER/WISCONSIN PUBLIC RADIO

values matter more than sophisticated production techniques.[12] Crane summarized those values as having respect for intelligence, having respect for each other, and having an incessant curiosity about everything. These values, he said, characterized the staff he found in Wisconsin.[13] He said he was surprised, as an outsider, how much the Wisconsin Idea continues to energize the staff.

Beyond their modifications in programming, Crane and Arnold, like Corriveau before them, dealt with the challenge and the opportunity of the Internet and mobile devices, which threatened all broadcast technology while providing new opportunities for service. Already a prolific producer of high-quality content and financially stronger than many public radio operations, WPR found itself in a better position than most to meet the challenges and exploit the opportunities. During Corriveau's years, WPR got serious about its online presence, largely due to the single-handed (and part-time) efforts of engineer Allen Rieland, "WPR's contemporary Earle Terry."[14] It was clear that Internet streaming and on-demand access to programming were central to radio's future, even if they never replaced broadcasting. WPR's most popular national program, *Whad Ya*

Morning classical music host Stephanie Elkins also hosts folk music on Sunday evenings. In 2012, she interviewed Gary Louris of the Jayhawks. JEFFREY POTTER/ WISCONSIN PUBLIC RADIO

Know?, provided a big head start in attracting audiences to its website, notmuch.com. WPR's investment in news reporters throughout the state combined with NPR news reporting provided the core for a strong on-demand news service on the station's site wpr.org.

More than any other programmer, however, Jean Feraca saw the trans-formational possibilities of multimedia digital streaming, podcasting, and on-demand programs. In 2003, with Schnirring's acquiescence, if not enthusiasm, she dropped her two-hour daily program to throw her heart and soul into a project she called *Here on Earth, Radio without Borders*. The entire Earth was her intended audience and her topics would know no borders. WPR broadcast the program and made it available to other sta-tions through the Public Broadcasting Exchange, but the program found few takers. While Feraca hoped some international broadcasters might carry the program, she understood that her main access to the world was through Internet streaming and podcasts. WPR could provide little budget for the project other than the salary of Feraca and her producer, Carmen Jackson, but Feraca's passion, charisma, and vision attracted some forty

students, interns, and volunteers eager to enhance international understanding on a basic human level and, not inconsequentially, contribute to
world peace. Feraca's interest in matters spiritual led her to concentrate on
understanding the values of Muslims. She won awards for her blog series
"Inside Islam: Dialogues and Debates" and for another series, "Muslims,
Mosques, and American Identity." Despite the high purpose and the hard
work, *Here on Earth* made little impact on a world scale or even in Wisconsin. Feraca had taken on an impossible, if noble, task. Discouraged, Feraca
decided to retire in 2012.

The better staffed and financed *To the Best of Our Knowledge* enjoyed
more success with its on-demand offerings, in part because it already
had developed a radio audience on stations across the country. A modest
"major gift" campaign to raise twenty-five thousand dollars to upgrade
TTBOOK's online presence succeeded quickly, a tribute to the program's
power and listener loyalty. In August 2015, *TTBOOK*'s podcast on author
David Foster Wallace ranked thirty-seventh on iTunes' podcast ranking.
Program director Mike Arnold observed, "WPR's twenty-five-year-old
program ranks higher right now than any of the new launches from

To the Best of Our Knowledge host Ann Strainchamps with producers Rehman Tungekar,
left, and Charles Monroe Kane. JEFFERY POTTER/WISCONSIN PUBLIC RADIO

companies with pithy names like Gimlet and Earwolf. WPR has just as much capacity to grow audience on new platforms as anybody . . . and we can do it with our mission and values intact."[15]

While digital technologies span the world, WPR's first commitment remains service to Wisconsin. The large volume of state content WPR produces provides a wealth of material for its wpr.org website. Following in the footsteps of NPR, which made its name synonymous with quality news, public radio stations across the country beefed up their news coverage partly to offset the decline in commercially supported journalism. WPR joined this public radio commitment to news, adding a second state capitol reporter and an investigative reporter to the already extensive statewide team in 2015. Approximately 150,000 listeners streamed or downloaded WPR programming each month that year. While most listeners still receive WPR content the old-fashioned way on their radios, the organization's "WPR Everywhere" initiative positions it to serve Wisconsin however audio media may evolve.[16]

One hundred years after it began as 9XM, Wisconsin Public Radio is a vital institution. It is grounded in the university that spawned it and valued by the citizens in the state whose name it bears. It has adapted the Wisconsin Idea in ways that Professor Terry could not have imagined but that would make him proud. It no longer has boundaries. It is available around the world, but it still shares its university's mission to search for truth. As Mike Arnold says, "We may be the first public radio entity to realize that it is actually a real asset to be affiliated with a university."[17] All those faces in the mural that still adorns the entry of Old Radio Hall would smile.

CONCLUSION

Revisiting the Wisconsin Idea

N ot long after I arrived in Madison, the *Capital Times* had asked, "Is the Wisconsin Idea dead?" While telling this story of the first hundred years of Wisconsin Public Radio and Wisconsin Public Television, I have made the case that the Wisconsin Idea still guides, to some degree, both public radio and public television in philosophy and practice. In a twenty-first-century media environment, they seek to embody Professor Lighty's original admonition "to interpret the true spirit, the life and the work of the university, as well as to instruct, stimulate, and enrich the lives of listeners."[1]

Nonetheless, the Walker administration's draft budget for 2015–2017 deleted "the Wisconsin Idea" from the University of Wisconsin mission statement. While public reaction caused the administration to quickly withdraw its language change, the proposal demonstrated that those in the governor's office saw the roles of government and of the university very differently than Robert La Follette, Charles McCarthy, Charles Van Hise, William Lighty, and their fellow early-twentieth-century progressives. The Walker administration proudly promoted business interests over progressive priorities like the university, public education, workers' rights, and environmental protection. Tax cuts prevailed over government services. Not surprisingly, therefore, the proposed budget backed away from state support for public radio and public television. WPR and WPT shared in the overall cut to the UW system under the governor's budget, but more seriously, the budget eliminated nearly all support for the ECB personnel who operated public television and radio transmitters across the state. Public broadcasting, the administration suggested, could raise its budget from private sources.

The ECB had long resisted the establishment of friends groups for public radio and television, but those groups now leapt into action to restore the cuts. They hired a lobbyist to make their case. Listeners and viewers bombarded their legislators with calls, letters, and e-mails. State Senator Luther Olsen convinced his fellow Republicans to reconsider the governor's proposal. Attacking services valued by hundreds of thousands of Wisconsinites proved to be bad politics. Moreover, the same towers and interconnection system that carried public broadcasting around the state carried emergency weather warnings and Amber Alerts. Endangering public safety was also bad politics. So legislators turned to a target with a less vocal constituency, the ECB's services to schools. However much K-12 educators might value what the ECB provided, losing those services was a relatively minor hit compared to those that recent state budgets had inflicted on public education. Educators did not rise to defend their ECB service, and the legislature axed programming for kids in school, once the primary rationale for state support for Wisconsin's educational radio and television networks. The legislature's decision left public service as the sole focus of public radio and television in Wisconsin.

The debate over state support for public broadcasting dramatized two competing positions on public broadcasting. The Wisconsin Idea envisioned public radio and television as a tax-supported service that served public purposes, such as enhancing democracy, building communities, stimulating minds, and providing important information. What we might call the Reagan Idea saw public broadcasting as one of many alternatives available to individuals to satisfy personal tastes and interests. This individualistic view suggests that those who enjoy such services should pay for them. Public broadcasters across the country mix these philosophies to varying degrees. Public broadcasters in the home state of the Wisconsin Idea give priority to programming for public purposes.

As I see it, current and emerging technologies are likely to push all media in the direction of the Reagan philosophy. Cable and satellite have provided individuals with many more choices than they had in the era of a few broadcasters reaching mass audiences. The Internet and mobile devices provide nearly infinite options. The idea of force-feeding audiences material they do not choose to view or hear is virtually dead. That idea had dominated when Britain's BBC began. The BBC maintained a monopoly that gave listeners and viewers no alternative to "good for you" programming perceived as being in the public interest. While educational radio and television never enjoyed such a monopoly in the United States, the founders of Wisconsin's noncommercial system saw the BBC as a model for their service.

Today, public broadcasting competes for attention in a world full of choices. Are any of those choices more worthy of taxpayer subsidies than others? Does a public interest override any of these individual preferences? If not, why shouldn't those who want a particular type of programming pay for it? This reasoning seems to apply to much that appears on public radio and television in Wisconsin. Public television's *Antiques Road Show* and *Downton Abbey* entertain select audiences who are no more entitled to a tax subsidy for their choices than those who pay for, or endure advertising that pays for, other tastes in entertainment. Public radio's *Car Talk, A Prairie Home Companion, Old Time Radio Night, Whad Ya Know?*, and *Wait, Wait, Don't Tell Me* are a great deal of fun, but not exactly public service. At one time, educational radio and television broadcast classical music and other high-culture offerings in the name of education. We know, however, that listeners and viewers today choose such offerings primarily for pleasure, relaxation, or inspiration—in other words, for the same reason people choose all types of music and entertainment.

The Wisconsin Idea, on the other hand, suggests that programming exists "to instruct, stimulate, and enrich the lives of listeners" and to connect listeners to the "life and work of the university." Today, the lectures of television's *University Place* and the interviews on radio's *University of the Air* remain classic educational programs that reach only small, if appreciative, audiences. They could not survive in a free-market media world and will always require subsidies.

I believe, however, that the Wisconsin Idea goes beyond narrow educational efforts like *University Place* and *University of the Air*. In this book, I have argued that the Wisconsin Idea provides an overall approach to programming that reflects the academic values of the University of Wisconsin. One can argue that the values of public broadcasting are as journalistic as they are academic, but fundamentally the two seek similar ends. Both journalism and universities search for truth by verifying facts, placing them in context, and debating them openly. Only then do they offer tentative, but seldom final, answers. Academics approach problems more slowly and rigorously than even the best journalists, but each seeks truth as best they can. Public broadcasting advances the outreach mission of universities when it applies those values to the exploration of issues and ideas for broad audiences. Professor Terry wanted 9XM to carry what he described as the university "atmosphere" to the boundaries of the state. He was not referencing specific information or formal courses so much as an overall approach to solving problems, attaining knowledge, and fostering culture. Substituting the word *values* for *atmosphere* in Terry's description captures the essence, and lasting importance, of public broadcasting in Wisconsin.

The values of the University of Wisconsin begin with the famous "sifting and winnowing" statement adopted by the Board of Regents on September 18, 1894, that paved the way for the Wisconsin Idea. "Whatever may be the limitations which trammel inquiry elsewhere," they wrote, "we believe that the great State University of Wisconsin should ever encourage that continual and fearless sifting and winnowing by which alone the truth may be found."[2] That search for truth takes place in the private intimacy of classrooms, offices, libraries, research labs, and, today, online. Publicly, it takes place on radio and television.

Whether in a campus building or on the public airwaves, the search for truth begins with a commitment to verifiable facts. "Sifting and winnowing" separates the most substantiated "facts" from less valid assertions and places the most pertinent facts into their historical, social, and moral context. The process requires sifting through all perspectives while winnowing out the weakest. For example, public broadcasters must seriously consider the views of those who deny global warming, but the winnowing process may find those views less convincing than the views of the vast majority of the scientific community.

This very public expression of objectivity, fairness, balance, and context in the search for truth remains the fundamental value of the Wisconsin Idea in public broadcasting. Today, digital technologies offer superior ways to reach specific audiences with specific content. For the general public, however, public broadcasting demonstrates, and thus promotes, the application of academic approaches to a full array of topics. With digital media delivering whatever an individual wants whenever he or she wants it, the role of public service broadcasting reverts to the goal of the founder of Britain's BBC, who promised to provide listeners with material they did not know they wanted. In an era when digital technologies allow a person to seek exactly what she or he wants to know—and only those views that will reinforce her or his existing biases and opinions—public broadcasting, in contrast, explores a variety of topics and viewpoints, some unexpected, some even unwanted. That approach, of course, is what higher education is all about. Whether considering health-care reform, tulip cultivation, or Victorian literature, public broadcasting ideally draws upon persons with real expertise—and makes them defend their positions. Its programs stimulate thought as they inform and even entertain.

While WPR's the Ideas Network connects quite clearly with the Wisconsin Idea, the programming on Wisconsin Public Television and radio's NPR News and Classical Network fill most of their hours with programming from national or international sources. Still, most programs from PBS or NPR reflect the values of the Wisconsin Idea. That idea shaped the thinking of NPR's founding philosopher, William Siemering. It shaped my thinking, too, as the first producer of NPR's seminal program, *All Things Considered*. NPR's style is tighter and its presentation slicker than Ideas Network programs, but NPR thrives because of its academic values. It appeals most strongly to those with the most education, those whose values were most shaped by their experiences in universities. Public television news and public affairs staples like *The News Hour, Frontline,* and *Washington Week in Review* have the same appeal. No one treats science more responsibly than PBS does on *Nova* and its numerous other nature programs. Even when Wisconsin Public Radio and Wisconsin Public Television broadcast these national programs, they serve the Wisconsin Idea.

Those who worked in educational radio and television tried to shed their "educational" label when they began to call themselves "public broadcasting" in the 1960s, but they failed. Public broadcasting was born in and nurtured by universities, and public broadcasters think like academics. Conservative critics who accuse public broadcasting of a "liberal bias" are often the same people who accuse universities, and particularly the university in Madison, of "liberal bias." Bias of any variety has no legitimate place in academia or in public broadcasting, although it undoubtedly exists, as it does in any human institution. Whether or not critics choose to believe it, both academics and public broadcasters strive to approach questions with open minds and produce answers that grow from verified facts. Both sometimes fall short, but no other institutions try harder.

Wisconsin Public Radio and Wisconsin Public Television cannot escape their origins in the Wisconsin Idea. Do they provide a genuine service to the state's citizens over and above the pleasure they provide to specific individuals? Your answer to that question will depend on your attitude toward the university and higher education in general. If you value one, you will value the other.

NOTES

Introduction: The Wisconsin Idea

1. David Pritchard, "The Selling of WHA," *The Capital Times* (Madison), December 15, 1978.
2. Pritchard, "The Selling of WHA," *The Capital Times* (Madison), December 16, 1978.
3. Pritchard, December 16.
4. Pritchard, December 15.
5. Charles McCarthy, *The Wisconsin Idea* (New York: MacMillan, 1912).
6. Ibid., 11.
7. Ibid., vii.
8. Ibid., 16.
9. "History of the Wisconsin Idea," University of Washington–Madison, http://wisconsinidea.wisc.edu/history-of-the-wisconsin-idea.
10. Roger Axford, research papers for dissertation, "William Henry Lighty, Adult Education Pioneer," Wisconsin Historical Society Archives, Madison, WI.
11. McCarthy, *The Wisconsin Idea*, 132–133.
12. Ibid.
13. Ibid., 303.
14. Ibid.
15. "Commercialism and Journalism," *Bulletin of the University of Wisconsin, General Series #386* (1913), Wisconsin Historical Society, General Collections, Madison, WI.
16. Hamilton Holt, "Can Commercial Journalism Make Good or Must We Look for the Endowed Newspaper," *Bulletin #386*.
17. Ibid.
18. President Calvin Coolidge, Address before the American Society of Newspaper Editors, Washington, DC, January 17, 1925.
19. "History of the Wisconsin Idea."

Chapter 1: Education Learns to Sing

1. Roger Penn, "The Origin and Development of Broadcasting at the University of Wisconsin to 1940" (PhD diss., University of Wisconsin, 1950), 128.
2. Ibid.
3. Mrs. I. F. Thompson to H. B. McCarty, May 2, 1949, quoted in Penn, "Origin and Development," 110.
4. Penn, "Origin and Development," 56.
5. Ibid., 52.
6. Ibid., 46.
7. Some of those tubes are still intact and on display in Vilas Hall.
8. Penn, "Origin and Development," 35.
9. William Lighty to Malcolm Hanson, n.d., in Roger Axford, research papers for dissertation, "William Henry Lighty, Adult Education Pioneer," Wisconsin Historical Society Archives, Madison, WI.
10. Andrew Hopkins, interview with Roger Axford, June 10, 1958, Axford Papers.
11. Larry Meiller, interview with author, May 21, 2014, Madison, WI.
12. William Lighty Papers, Wisconsin Historical Society Archives, Madison, WI.
13. Axford Papers.
14. William Lighty, conference paper written for the 8th Annual Convention of the National University Extension Association in St. Louis, 1923, Lighty Papers.
15. Penn, "Origin and Development," 261.
16. Roger Axford, *William Henry Lighty, Adult Education Pioneer* (Chicago: University of Chicago Press, 1961).
17. C. H. Alzmeyer to Earle Terry, 1925, WHA General Correspondence File, University of Wisconsin Archives, Madison, WI (hereafter cited as WHA GCF).
18. Ibid.
19. Hopkins interview.
20. Ibid.
21. Axford Papers.
22. Ibid.
23. Letter to Professor Terry, box 2, folder 7, October 29, 1927, WHA General

Subjects File, University of Wisconsin Archives, Madison, WI (hereafter cited as WHA GSF).

24. Letter to H. B. McCarty, August 29, 1932, box 2, folder 12, WHA GSF.

25. H. R. Doering to H. B. McCarty, May 19, 1932, box 2, folder 12, WHA GSF.

26. Penn, "Origin and Development," 182.

27. Ibid.

28. Henry Ewbank, quoted in Penn, "Origin and Development," 54.

29. Ibid.

30. Penn, "Origin and Development," 187.

31. Letter to Earle Terry, July 5, 1922, quoted in Axford dissertation.

32. Axford Papers.

33. Hopkins interview.

34. Henry Ewbank and Jeffery Aurer, *Discussion and Debate: Tools for Democracy* (New York: F.S. Crofts and Company, 1941).

35. Henry Ewbank Papers, University of Wisconsin Archives, Madison, WI.

36. Radio Committee to President Glenn Frank, December 24, 1928, WHA GCF.

37. Helen Matheson, profile on Henry Ewbank, *Wisconsin State Journal*, August 17, 1947.

38. Ibid.

39. Glenn Frank, "Radio as an Educational Force," *Annals of the American Academy of Political and Social Science* 177 (January 1935): 119–122.

40. Ibid.

41. Ibid.

Chapter 2: The State Stations

1. Roger Penn, "The Origin and Development of Broadcasting at the University of Wisconsin to 1940" (PhD diss., University of Wisconsin, 1950), 35.

2. Karl Schmidt, interview with author, June 6, 2013, Madison, WI.

3. H. B. McCarty to listener, n.d., WHA General Correspondence File, University of Wisconsin Archives, Madison, WI (hereafter cited as WHA GCF).

4. Schmidt interview.

5. Harold Engel Papers, Wisconsin Historical Society Archives, Madison, WI.

6. Ibid.

7. Glenn Frank, "Education by Radio," 1932, box 9, folder 15, WHA GCF.

8. Engel Papers.

9. "The British Broadcasting Corporation," Britannica.com.

10. Harold Engel to Reverend Father Charles E. Coughlin, December 1932, box 2, folder 12, WHA GCF.

11. Ibid.

12. Engel Papers.

13. Ibid.

14. Penn, "Origin and Development," 402.

15. Ibid.

16. G. E. Mendenhall to Henry Ewbank, January 14, 1932, box 2, folder 12, WHA GCF.

17. Engel Papers.

18. Extension Service of the College of Agriculture, "WHA Broadcasts Farm Programs," box 15, folder 24, WHA General Subjects File, University of Wisconsin Archives, Madison, WI (hereafter cited as WHA GSF).

19. Extension Service of the College of Agriculture, "WHA Broadcasts Home-maker Programs," box 2, folder 19, WHA GSF.

20. Erika Janik, "Dear Mrs. Hazard," *On Wisconsin* (Spring 2007): 41.

21. Ibid.

22. Ibid., 43.

23. Engel Papers.

24. Committee on Radio Broadcasting, "Total Enrollment by Course of Wisconsin College of the Air, 1934–1935," box 2, folder 19, WHA GSF.

25. WHA Radio, on-air announcement, box 2, folder 10, WHA GSF.

26. Technically, the BBC always remained independent of the government that created and funded it.

27. Remarks of H. B. McCarty at retirement of Fannie Steve, May 1966, box 2, folder 28, WHA GSF.

28. Wisconsin School of the Air, script for "Afield with Ranger Mac," November 16, 1953, and stationery for Ranger Mac's Conservation Club, box 22, folder 10, WHA GSF.

29. University Radio Committee, "1945 Annual Report," box 2, folder 8, WHA GSF.

30. Ibid.

31. Research Project in School Broadcasting, "Second Year Report, 1938–39," box 26, folder 14, WHA GSF.

32. Ibid.

33. Ibid.

34. Josh Sheppard, "Broadcasting at the Office of Education, 1934–1944" (PhD diss., University of Wisconsin, 2013), 223.

35. Ibid.

36. Engel Papers.

37. Professor Henry Ewbank to Governor Phillip La Follette, July 29, 1932, box 2, folder 18, WHA GSF.

38. Ibid.

39. Ibid.

40. Penn, "Origin and Development," 363.

41. Engel Papers.

42. Harold Engel, "Agreement Covering the Use of Stations WHA and WLBL in the Wisconsin Pre-Primary Campaign," August 12, 1932, box 15, folder 12, WHA GSF.

43. Ibid.

44. Ibid.

45. Engel Papers.

46. Ibid.

47. Schmidt interview.

48. Engel Papers.

49. Ibid.

50. Henry Ewbank, response to League of Wisconsin Radio Stations, October 1948, Engel Papers, WHS Archives.

51. Charles Hill to H. B. McCarty, October 20, 1937, box 3, folder 2, WHA GSF.

52. Charles Siepmann, "Review of Wisconsin," handwritten notes, April 1–June 9, 1937, Rockefeller Foundation Archives, Westchester County, NY.

53. Ibid.

54. Schmidt interview.

55. Siepmann, "Review of Wisconsin."

Chapter 3: The State Radio Network

1. H. B. McCarty and Harold Engel, interview with Steven Lowe, Madison, 1972, University Archives Oral History Project, University of Wisconsin Archives, Madison, WI.

2. H. B. McCarty, untitled, June 14, 1946, *Journal of FM*, box 11, folder 1, WHA General Subjects File, University of Wisconsin Archives, Madison, WI (hereafter cited as WHA GSF).

3. State Radio Council of Wisconsin, "Statement of Policy," December 6, 1938, box 27, folder 15, WHA GSF.

4. Ibid.

5. H. B. McCarty to Henry Ewbank, May 1943, Henry Ewbank Papers, University of Wisconsin Archives, Madison, WI.

6. Karl Schmidt, interview with author, June 6, 2013, Madison, WI.

7. University Radio Committee, Annual Report, 1945, box 3, folder 8, WHA GSF.

8. Scripts for seven "programs for wartime," box 31, folder 6, WHA GSF.

9. Henry Ewbank Papers, box 3.

10. State of Wisconsin in Assembly, No. 631, A, May 25, 1945: A bill to create 20.143 and 43.60 of the statutes, relating to the creation of a state radio council, the establishment of a state broadcasting system for educational purposes and making an appropriation, box 10, folder 16, WHA GSF.

11. H. B. McCarty to E. B. Fred, March 9, 1945, box 1, folder 17, WHA GSF.

12. Ibid.

13. University Radio Committee, "Further Statement of Policy Concerning Controversial Issues over Station WHA," December 11, 1945, box 3, folder 8, WHA GSF.

14. Ibid.

15. Schmidt interview.

16. State of Wisconsin in Assembly, No. 631, A.

17. Helen Stanley, script for "WHA-FM Inaugural Broadcast," 2:00–2:30 p.m., March 30, 1947, box 10, folder 14, WHA GSF.

18. Ibid.

19. Phil Drotning, "Under the Dome: Some Hot Potatoes for Chef Rennebohm," July 19, 1949, *Wisconsin State Journal* (Madison), box 10, folder 18, WHA GSF.

20. The letters were sorted by counties and fill box 11, WHA GSF.

21. Harold Engel Papers, Wisconsin Historical Society Archives, Madison, WI.

22. H. B. McCarty, personal note to Governor Oscar Rennebohm, August 4, 1949, box 10, folder 18, WHA GSF.

23. Engel Papers.

24. Letters filed in box 10, WHA GSF.

25. Ewbank Papers.

26. Ibid.

27. H. B. McCarty, untitled, June 14, 1946, *Journal of FM*, WHA GSF.

28. State Radio Council, "Review of Educational Broadcasting in Wisconsin," September 15, 1952, box 12, folder 8, WHA GSF.

29. Linda Clauder, interview with author, September 2013, Madison, WI.

30. H. B. McCarty, "Prospectus on Programming for the Proposed State FM Network," box 11, folder 1, WHA GSF.

31. Ewbank Papers, box 3.

32. Ibid., box 4.

33. H. B. McCarty, handwritten note on memo to him from Roy Vogelman, April 1, 1963, box 12, folder 1, WHA GSF.

34. Schmidt interview.

35. University Radio Committee, Annual Report, January 13, 1947, box 3, folder 26, WHA GSF.

36. Wisconsin State Broadcasting Service, "Music in Context," n.d., brochure from personal collection of Linda Clauder.

37. Ibid.

38. Clauder interview.

39. Tom Clark, interview with author, September 2, 2013, Madison, WI.

40. University Radio Committee, Annual Report, 1947.

41. Grant Fairley, *Look Up—Way Up: The Friendly Giant; The Biography of Robert Homme* (Toronto: Palantir Publishing, 2007).

42. *WHA Announcers Manual*, n.d., compiled by William G. Harley, box 1, folder 10, WHA GSF.

43. Karl Schmidt to Mel Carlson, July 24, 1964, box 7, folder 1, WHA General Correspondence File, University of Wisconsin Archives, Madison, WI.

44. Clauder interview; Clark interview; Larry Meiller, interview with author, May 21, 2014, Madison, WI; Jim Fleming interview with author, May 2014, Madison, WI.

45. Fleming interview.

46. Schmidt, Clark, Clauder, Fleming, and Meiller interviews.

Chapter 4: A State Television Network?

1. Report of the National Association of Educational Broadcasters, Educational Television Seminar, June 21–27, 1953, Harold Engel Papers, Wisconsin Historical Society Archives,Madison, WI.
2. Robert Blakely, *To Serve the Public Interest: Educational Broadcasting in the United States* (Syracuse, NY: Syracuse University Press, 1979), 83.
3. Blakely, *To Serve the Public Interest*, 87.
4. Ford Foundation to the University of Wisconsin Board of Regents, December 18, 1953, Henry Ewbank Papers, Wisconsin Historical Society Archives, Madison, WI.
5. "WHA-TV Goes on Air on May 3. Programs Will Reach 15 Miles," *Wisconsin State Journal* (Madison), April 25, 1954.
6. Boris Frank, interview with author, June 4, 2013, Madison, WI.
7. UW Extension, 1965 Annual Report, box 30, folder 1, Radio/Television Division Files, Don McNeil Papers, University of Wisconsin Archives, Madison, WI.
8. Program guide, Box 7, folder 14, WHA General Subjects File, University of Wisconsin Archives, Madison, WI (hereafter cited as WHA GSF).
9. Robert Homme qtd. in Grant Fairley, *Look Up–Way Up: The Friendly Giant; The Biography of Robert Homme* (Toronto: Palantir Publishing, 2007).
10. Ibid.
11. Wisconsin State Radio Council, press release, April 16, 1956, box 11, folder 10, WHA GSF.
12. Wisconsin State Radio Council, undated press release, "CBC Praises Friendly Giant," box 11, folder 10, WHA GSF.
13. University Radio-Television Committee, 1954 Annual Report (February 1955), box 3, folder 16, WHA GSF.
14. H. B. McCarty, "History of the State Radio Council," April 12, 1966, McNeil Papers.
15. Blakely, *To Serve the Public Interest*, 85.
16. Harold Engel Papers, Wisconsin Historical Society Archives, Madison, WI.
17. Ibid.
18. Ibid.
19. State Radio Council, 1953 Annual Report (February 1954), box 3, WHA GSF.

20. Mike McCauley, "The Battle over Educational TV in Wisconsin: 1952–1954" (unpublished paper), 7.

21. McCauley, "The Battle over Educational TV," 8.

22. Engel Papers.

23. McCauley, "The Battle over Educational TV," 11.

24. Editorial, *Milwaukee Journal*, October 19, 1954, box 6, folder 13, WHA GSF.

25. Advertisement, *Algoma Record*, October 1954, box 6, folder 11, WHA GSF.

26. Ibid.

27. Harold Engel to Wisconsin Committee on State-Owned, Tax-Supported Television, ca. 1954, box 31, folder 14, WHA GSF.

28. Engel Papers.

29. Ibid.

30. McCauley, *To Serve the Public Interest*, 11.

31. Karl Schmidt, interview with author, June 6, 2013, Madison, WI.

32. State Radio Council Statement, October 20, 1954, box 6, folder 14, WHA GSF.

33. "State ETV Proposal Loses by 2-1," *Milwaukee Sentinel*, November 4, 1954.

34. Ewbank Papers.

35. Boris Frank, interview with author, June 4, 2013, Madison, WI.

36. McNeil Papers, box 30.

37. Harold B. McCarty, "Memo from HBM to LSD, 12/26/63," box 6, folder 14, WHA GCF.

38. Ibid.

39. Ibid.

40. Editorial, *The Capital Times* (Madison, WI), April 1966.

41. Karl Schmidt to Mel Carlson, July 24, 1964, box 7, folder 1, WHA GCF.

42. Robben Fleming was the father of future Wisconsin Public Radio personality Jim Fleming.

43. Box 13, McNeil Papers.

44. Ibid.

Chapter 5: UW Extension

1. Karl Schmidt, interview with author, June 6, 2013, Madison, WI.

2. Ibid.

3. H. B. McCarty, 1966–67 Report on Radio and Television for University

Extension, June 30, 1967, box 30, Radio/Television Division Files, Don McNeil Papers, University of Wisconsin Archives, Madison, WI.

4. Jim Robertson, memo to Don McNeil, April 1967, box 13, McNeil Papers.
5. Schmidt interview.
6. General Extension policy statement, March 31, 1954, box 2, Henry Ewbank Papers, University of Wisconsin Archives, Madison, WI.
7. Memo from Jim Collins to Karl Schmidt, August 1968, box 8, folder 4, WHA General Correspondence File, University of Wisconsin Archives, Madison, WI (hereafter cited as WHA GCF).
8. Glen Pulver to Jim Robertson, June 18, 1968, box 8, folder 14, WHA GCF.
9. Ron Bornstein and Jim Robertson to Don McNeil, May 15, 1967, box 30, McNeil Papers.
10. Schmidt interview.
11. Ralph Johnson and Ron Bornstein, manuscript for "The Inner Core: City within a City," *National Association of Educational Broadcasters Journal*, November 1968, box 47, McNeil Papers.
12. Ibid.
13. Karl Schmidt, memo to radio staff, September 1, 1967, box 7, folder 6, WHA GCF.
14. Ibid.
15. Johnson and Bornstein, "The Inner Core," 3.
16. Jim Robertson to Don McNeil, February 2, 1968, box 47, folder 1, McNeil Papers.
17. Governor Knowles statement, February 22, 1968, box 47, folder 1, McNeil Papers.
18. Schmidt interview.
19. Note from John Macy, April 15, 1969, box 8, WHA GCF.
20. Johnson and Bornstein, "The Inner Core," 6.
21. Ibid.
22. Ibid.
23. Karl Schmidt to Don McNeil, May 1968, box 8, folder 14, WHA GCF.
24. Report of conversation with Roger LaGrand, May 10, 1968, box 8, folder 14, WHA GCF.
25. Glen Pulver to Jim Robertson, June 18, 1968, box 8, folder 14, WHA GCF.
26. Boris Frank, interview with author, June 4, 2013, Madison, WI.
27. WHA-TV application to Commission on Aging, box 29, folder 3, WHA

General Subjects File, University of Wisconsin Archives, Madison, WI (hereafter cited as WHA GSF).

28. Ibid.

29. Frank interview.

30. Boris Frank to Dick Lutz, July 31, 1968, box 8, folder 8, WHA GCF.

31. *Project 360* Final Report, 1972–1973, box 20, folder 3, WHA GSF.

32. UW Extension, Nutrition Committee comments, September 10, 1971, box 26, folder 5, WHA GCF.

33. *RFD* rundown, May 7, 1971, box 26, folder 1, WHA GCF.

34. Report from Boris Frank on meeting with US Office of Education, May 12, 1971, box 26, folder 1, WHA GCF.

35. Ibid.

36. Script for 360 Radio Program #1, box 20, folder 2, WHA GSF.

37. Janice C. Wheaton to Tony Tiano, March 6, 1973, box 20, folder 2, WHA GSF.

38. Frank interview.

39. *American Pie Forum* proposal, box 1, folder 7, WHA GSF.

40. *American Pie Forum* scripts, box 1, folder 9, WHA GSF.

41. *American Pie Forum* proposal.

42. Progress report on *American Pie Forum*, December 1973, box 8, WHA GCF.

43. Show One script, *American Pie Forum*, box 29, folder 2, WHA GSF.

44. *American Pie Forum* evaluations, box 1, folder 8, WHA GSF.

45. Progress report on *American Pie Forum*.

46. James Killian Jr., *Public Television: A Program for Action; A Report of the Carnegie Commission on Educational Television* (New York: Bantam Books, 1967).

Chapter 6: Public Television

1. Robert Blakely, *To Serve the Public Interest: Educational Broadcasting in the United States* (Syracuse, NY: Syracuse University Press, 1979), 172.

2. Ibid., 178.

3. Ibid., 182.

4. Carnegie Commission, *Public Television: A Program for Action; A Report of the Carnegie Commission on Educational Television* (New York: Bantam Books, 1967).

5. Ibid.

6. Boris Frank, interview with author, June 4, 2013, Madison, WI.

7. Undated memo to Chancellor McNeil, provided by Boris Frank.

8. Ibid.

9. Dick Lutz to Bernard Cherin, December 16, 1969, box 4, folder 8, WHA General Correspondence File, University of Wisconsin Archives, Madison, WI (hereafter cited as WHA GCF).

10. Ibid.

11. Dick Lutz, Proposal to Ford Foundation, n.d., box 26, folder 24, WHA General Subjects File, University of Wisconsin Archives, Madison, WI (hereafter cited as WHA GSF).

12. Owen Coyle, Report to Ford Foundation, n.d., box 26, folder 22, WHA GSF.

13. Dick Lutz, speech to Madison Kiwanis, April 17, 1970, box 27, folder 1, WHA GSF.

14. *Six30* file, box 26, folder 24, WHA GSF.

15. Ibid.

16. Ibid.

17. Report of Extension University Committee, April 6, 1970, box 27, folder 2, WHA GSF.

18. David Cronon to Dick Lutz, box 27, folder 6, WHA GSF.

19. William Wedemeyer to Luke Lamb, box 27, folder 7, WHA GSF.

20. Report of Extension University Committee, April 6, 1970.

21. Ibid.

22. Dick Lutz, mock press release, July 1, 1970, box 26, folder 20, WHA GSF.

23. Paul Norton, informal conversations with author, 1981–1984, Madison, WI.

24. John Wyngaard, "Educational TV Holds Threats in Its Promises," *Wisconsin State Journal* (Madison), December 29, 1968.

25. Jim Robertson to Don McNeil, November 26, 1966, box 3, folder 2, McNeil Papers, University of Wisconsin Archives, Madison, WI.

26. Jim Robertson to CCHE, April 7, 1967, box 3, folder 2, McNeil Papers.

27. Ibid.

28. Governor Warren Knowles to members of the Educational Communications Board, January 23, 1968, box 42, folder 10, McNeil Papers.

29. Ibid.

30. Lee Sherman Dreyfus to CCHE, April 22, 1966, box 3, folder 10, McNeil Papers.
31. Ron Bornstein, "Four Statewide TV Issues," 1971, box 26, folder 12, WHA GSF.
32. Ibid.
33. Ibid.
34. Jim Robertson to Henry Algren, September 1968, box 42, folder 10, McNeil Papers.
35. Byron Knight, interview with author, January 14, 2014, Middleton, WI.
36. Wilson Thiede to President John Weaver, February 26, 1976, Educational Communications Board archives, Wisconsin Public Broadcasting Center, Madison, WI (hereafter cited as ECB archives).
37. Larry Dickerson, interview with author, November 15, 2013, Madison, WI.
38. Robert Dries, "Commercial Television Is Casting Wary Eyes on WHA," *Madison Magazine*, January 1983, ECB archives.
39. Ibid.
40. Bornstein memo to staff, January 5, 1983, ECB archives.
41. "The War at Home," *New York Times*, http://www.nytimes.com/movies/movie/52754/The-War-at-Home/awards.
42. Byron Knight, interview with author, March 17, 2014, Madison, WI.
43. Larry Dickerson, interview with author, November 15, 2013, Madison, WI.
44. Knight interview, March 17.
45. Knight interview, January 15.
46. Malcolm Brett, interview with author, March 24, 2014, Madison, WI.
47. Dan Peterson to Mick Rhodes, "Search for the Violin," September 14, 1979, box 31, folder 4, WHA GSF.
48. Friends of WHA History, unpublished manuscript, author unknown, 1984, Office of the Director of Broadcasting and Media Innovations, Vilas Hall, Madison, WI.
49. Ibid.
50. Ibid
51. Brett interview.
52. *Tryout TV* evaluations, box 30, folder 10, WHA GSF.
53. Ibid.
54. Friends of WHA History.
55. Dickerson interview.

56. *New Tech Times* file, box 14, folder 1, WHA GSF.

57. Brett interview.

58. *New Tech Times* file.

59. Richard Hiner to Tony Moe, March 30, 1981, ECB archives.

60. Brett interview.

Chapter 7: And Radio

1. Karl Schmidt, memo to staff, September 29, 1967, box 7, folder 4, WHA General Correspondence File, University of Wisconsin Archives, Madison, WI (hereafter cited as WHA GCF).

2. Fred Harvey Harrington, congressional testimony, April 14, 1967, box 30, Radio/Television Division Files, Don McNeil Papers, University of Wisconsin Archives, Madison, WI.

3. Ibid.

4. Karl Schmidt, Reaction to NAEB radio planning meeting in Chicago, January 1964, box 6, folder 14, WHA GCF.

5. Karl Schmidt, interview with author, June 6, 2013, Madison, WI.

6. Ibid.

7. James Brown, "Untitled," *LA Times*, November 20, 1977, box 5, folder 4, WHA General Subjects File, University of Wisconsin Archives, Madison, WI.

8. Schmidt interview.

9. Ibid.

10. Karl Schmidt, memo to staff, October 30, 1967, box 7, folder 6, WHA GCF.

11. Joe Grant, memo to Ralph Johnson, November 5, 1970, box 14, folder 22, WHA GCF.

12. Jim Fleming, interview with author, May 2014, Madison, WI.

13. Joe Grant, memo to Ralph Johnson, February 24, 1971, box 14, folder 22, WHA CGF.

14. Ralph Johnson to "To Whom It May Concern," February 25, 1971, box 14, folder 22, WHA GCF.

15. Claire Kentzler, telephone conversation with author, spring 1974.

Chapter 8: Wisconsin Public Radio

1. Larry Dickerson, "Restructuring Public Radio and Television in Wisconsin," p. 18, unpublished study, June 2000, Educational Communications

Board archives, Wisconsin Public Broadcasting Center, Madison, WI (hereafter cited as ECB archives).

2. Jim Fleming, interview with author, May 2014, Madison, WI.

3. Robert Blakely, *To Serve the Public Interest: Educational Broadcasting in the United States* (Syracuse, NY: Syracuse University Press, 1979).

4. William Siemering, NPR statement of purpose, 1970, available at http://current.org/2012/05/national-public-radio-purposes.

5. Joy Cardin, interview with author, May 15, 2014, Madison, WI.

6. Larry Meiller, interview with author, May 21, 2014, Madison, WI.

7. ECB archives.

8. "Operating Policy for Radio and Television," Endorsed by Council of Chancellors, September 29, 1967, box 30, Radio/Television Division Files, Don McNeil Papers, University of Wisconsin Archives, Madison, WI.

9. Jack Mitchell, "A Plan for Radio One/Radio Two in Wisconsin," undated proposal, ECB archives.

10. Ibid.

11. Roger Gribble, "Central Radio Plan Gets Cold Reception," *Wisconsin State Journal* (Madison), February 7, 1981.

12. Ibid.

13. Ibid.

14. Mordecai Lee to Co-chairs of Joint Legislative Audit Committee, February 15, 1983, ECB archives.

15. Legislative Audit Bureau, Report 83-30, An Evaluation of Public Radio Stations Serving Milwaukee, October 20, 1983, ECB archives.

16. Ibid.

17. Dickerson, "Reconstructing Public Radio and Television in Wisconsin," 19.

18. Joint press release from three stations, WHAD, WUWM, and WYMS, February 22, 1983, ECB archives.

19. Mike Zahn, "News from WUWM Bad News to WHAD," *The Milwaukee Journal*, November 8, 1987.

20. Memo from Paul Norton to Jack Mitchell, "Proposed Reorganization of Wisconsin Public Radio," November 23, 1983, ECB archives.

Chapter 9: Wisconsin Public Television

1. Byron Knight, interview with author, January 14, 2014, Madison, WI.

2. Richard Lawson, memo to WHA staff, February 8, 1989, Educational

Communications Board archives, Wisconsin Public Broadcasting Center, Madison, WI (hereafter cited as ECB archives).

3. Paul Norton, memo to ECB staff, April 21, 1989, ECB archives.

4. Ibid.

5. Ibid.

6. Knight interview.

7. James Steinbach, interview with author, April 1, 2014, Madison, WI.

8. Byron Knight, memo to author, January 2014.

9. James Steinbach, e-mail to author, April 14, 2014.

10. Steinbach interview.

11. Knight memo.

12. Dave Iverson, telephone interview with author, March 12, 2014.

13. Ibid.

14. Steinbach interview.

15. Knight interview.

16. Andy Moore, interview with author, April 3, 2014, Madison, WI.

17. Ibid.

18. Ibid.

19. Larry Dickerson, interview with author, November 15, 2013, Madison, WI.

20. Iverson interview.

21. Jerry Huffman, interview with author, April 10, 2014, Madison, WI.

22. Ibid.

23. Ibid.

24. Moore interview.

25. Steinbach interview.

26. Iverson interview.

27. Moore interview.

28. Ibid.

29. Huffman interview.

30. Moore interview.

31. Ibid.

32. Iverson interview.

33. Malcolm Brett, interview with author, March 24, 2014, Madison, WI.

34. Dickerson interview.

35. Knight interview.

36. Kathy Bissen, interview with author, June 13, 2014, Madison, WI.

37. Brett interview.

38. Bissen interview.

39. See the Wisconsin Hometown Stories website, http://wpt.org/Wisconsin-Hometown-Stories.

40. Steinbach interview.

41. See the Wisconsin World War II Stories website, http://wpt.org/Wisconsin-War-Stories/world-war-ii-stories/main.

42. Steinbach interview.

43. Ibid.

44. See the LZ Lambeau Sponsors webpage, http://wpt.org/lzlambeau/sponsors.

45. Steinbach interview.

46. National Center for Media Engagement website, http://mediaengage.tumblr.com/.

47. Bissen interview.

48. Michael Harryman, e-mail to author, October 2015.

49. Byron Knight, memo to author, January 2014.

Chapter 10: The Ideas Network

1. Joy Cardin, interview with author, May 15, 2014, Madison, WI.

2. Ibid.

3. Greg Schnirring, interview with author, June 25, 2014, Madison, WI.

4. Ibid.

5. Ibid.

6. Dean Kallenbach, telephone interview with author, July 1, 2014.

7. Schnirring interview.

8. Cardin interview.

9. Mike Crane, interview with author, July 3, 2014, Madison, WI.

10. Rebecca Dopart, e-mail to WPR staff, July 28, 2014, WPR email archives, Madison, WI.

11. Cardin interview.

12. Mike Arnold, interview with author, July 1, 2014, Madison, WI.

13. Mike Crane interview

14. Ibid.

15. Mike Arnold, e-mail to WPR staff, August 14, 2015, WPR email archives.

16. Jeffrey Potter, e-mail to author, November 9, 2015.
17. Arnold interview.

Conclusion: Revisiting the Wisconsin Idea

1. Quoted in Roger Penn, "The Origin and Development of Broadcasting at the University of Wisconsin to 1940" (PhD diss., University of Wisconsin, 1950), 261.
2. Board of Regents, quoted in Charles McCarthy, *The Wisconsin Idea* (New York: McMillan, 1912), 28–29.

INDEX

Page numbers in *italics* indicate illustrations.

9XM, 12, 13, 14, 33
  See also WHA
360 (TV show), 90–92
970 Radio Association, 141, 142

About Cars, 181
Accent on Living, 127
adult education, 14, 90–94
advertisements, 134–135, 182, 183
Agard, Walter, 51
agricultural extension, 4, 79, 81
agricultural programming, 15, 27–28,
  33, 36, 45, 59
Ahlgren, Henry, 105
"Aims of Educational Broadcasting"
  (document), 55
Airwaves, 118–119
Albee, Edward, 124, 125
All Things Considered, 121
AM Saturday, 139
American Ism, An, 112
American Pie Forum, 90–94, *93*
Andreasen, Margaret, 136–137, *137*
*Annals of the American Academy of
  Political and Social Science*, 23
Arnold, Mike, 183, 184, 185, 187–188
art programs, 36, 39–40
athletic events. *See* sports programs
audiences
  listeners of Wisconsin Public
    Radio in Madison, 176
  participation in programming
    choices, 161, 190
  of public broadcasting, 100
  reaction to *American Pie Forum*, 94
  reaction to call-in programming,
    138
  reaction to music programming
    changes, 178, 180
  reaction to inner core project, 89

reaction to *Out of Context*, 128–129
reaction to radio dramas, 124
reaction to *Six30*, 103–105
reaction to sports broadcasts, 19
rural, 15, 86, 90–94
size of, 56
support of educational radio and
  television, 73–76, 120
support of state FM network, 53,
  55
support of Wisconsin Public
  Radio and Wisconsin Public
  Television, 142–143, 168–173,
  190
urban, 86–89
automobile advice programming, 138,
  181
awards for programming, 112, 116,
  124

Bailey, George, 124, 144–145, 176
Bandwagon, The, 60–62
*Bandwagon Correspondence School,
  The*, 60–62
Barkan, Ben, 88
Bartell, Gerald, 46, 47, 52
BBC. *See* British Broadcast
  Corporation (BBC)
Bennett, Edward, 12, 25, 28, 30, 33,
  34
Birge, Edward, 18, 21
Bissen, Kathy, 156, 171
Bliss, Milton, 33
Bornstein, Ron, *86*
  defense of local programming,
    108, 145
  director of WHA, 83, 85, 87, 89,
    100, 112, 115
  ECB relationship, 108, 112, 145,
    148

public radio, 122, 125, 127,
130–132
public television, 83, 85, 87, 89,
100, 155
support for "friends" groups, 115,
141
vice president, UW system, 119,
151, 180
Brett, Malcolm, 163, 165, 166, 167,
168
British Broadcast Corporation (BBC),
37, 42, 45, 46, 133, 190–191
broadcasting
digital, 163
educational vs. public, 99
as new technology, 7
Wisconsin Idea and, 7, 23–24,
78–79, 192
See also public broadcasting;
public radio; public television;
Wisconsin Idea; Wisconsin
Public Radio; Wisconsin
Public Television
Brown, Charles E., 17–18
Brown, James, 124
budgets. See funding

call-in programs, 135–138
Canadian Broadcasting Corporation
(CBC), 71
Canary, David, 92, 93–94
Car Talk, 138, 181
Cardin, Joy, 148–149, 181, 183, 184
Carnegie Commission on Educational
Television, 99–100, 116, 120
Central Time, 184
Chapter a Day, 45
Chautauqua tent shows, 16
children's programming, 37–41, 39,
68–71, 156, 166–167
Clark, Tom, 137
assistant program director, 64
news and talk show host, 136, 138,
177
on-air fund-raising, 142
retirement of, 183
classical music, 1, 16, 60, 112. See also
music programming

Clauder, Linda, 134
Clayton, Norman, 56
clear channel stations, 26–27, 30, 44,
48
Clodius, Robert, 79–80
College of the Air, 35–36, 37, 51, 56, 63
Collins, Jim, 62–63, 64, 84, 85, 130
commercial radio, 2, 19, 20, 28, 29,
30, 31, 133–134
commercial television, 74, 99–100,
112
"Commercialism and Journalism"
conference, 6
community engagement, 168–172,
176
continuing education. See UW
Extension; Wisconsin College of
the Air
"Controversial Issues During
Wartime" (document), 50
controversial programming
Ewbank's support for open
discussion, 50
Out of Context, 128–129
Selling of America, The, 116
Six30, 102–103, 105
Tongues Untied, 153
Tryout TV, 116
University Forum, 57–58
Conversations with Margaret
Andreasen, 137
cooperative extension. See
agricultural extension
Corporation for Public Broadcasting
(CPB)
creation of, 99
funding local stations, 110, 121,
142, 144
radio drama center created by, 122
Corrazzi, Joe, 92
Corriveau, Phil, 182, 185
Cotter, Carol, 114
Coughlin, Charles, 29
Couple Next Door, The, 123
Coyle, Owen ("Dick"), 102, 104
CPB. See Corporation for Public
Broadcasting (CPB)
Crane, Mike, 183, 185

Crisis in the City, 76
Cronon, David, 104
cultural programming, 16, 36, 49, 68, 168

Daniels, Sarah, *137*
Davison, Glenn, 181
debates
    broadcasts of, 42–43, 159–163
    importance to democracy, 22, 42, 43
    University Forum, 58
Dickerson, Larry, 111, 113, 116, 158, 164
digital broadcasting, 163
Director's Cut, 167
Discussion and Debate (Ewbank), 22
documentaries
    African Americans, 88, 153, 171–172
    anti-war demonstrations, 112, 128
    commercialism, 116
    Frank Lloyd Wright, 114
    Joe McCarthy, 112
    Marlon Riggs, 153
    Native Americans, 114
    segregation in Milwaukee housing, 88, 171–172
    social issues, 88, 128, 153, 171–172
    Vel Phillips, 171–172
    veterans' stories, 169–170
    Vietnam, 114
    violins, physics and history of, 114
    Wisconsin communities, 168
    See also names of individual programs
Doyle, Jim, 159
Dreyfus, Lee Sherman, 77–80, *78,* 83, 108
Dunn, Kathleen, 177–178, *179*
Dyke, Bill, 102–103

Earplay, 122–126
ECB. *See* Educational Communications Board (ECB)
educational broadcasting
    programming for, 28
    vs. public service broadcasting, 99, 107

state stations and, 28–29
structure of, 164–166
See also educational radio; educational television; state radio network; state stations; state television network; University of Wisconsin Extension; Wisconsin Public Radio; Wisconsin Public Television
Educational Communications Board (ECB)
    antagonism with university, 152
    budget cuts, 190
    creation of, 107
    Knight, program director, 112, 153–154, 165–166
    Mitchell, director of radio, 148–150
    Norton, executive director, 133, 139, 142–144, 147
    State Radio Council replaced by, 107
    state radio network and, 126
    studio in Vilas Hall, 110
    UW Extension staff, tensions with, 126, 147–148
    WHA-TV and, 108–110, 112, 113–114, 152
    Wisconsin Public Radio approved by, 133
    See also state television network; Wisconsin Public Television
educational radio
    channels reserved for, 48
    competition with commercial broadcasting, 20, 29
    democracy benefits from, 41–42
    evaluation of, 40–41
    goals for, 23
    programming of, 16–20
    purpose of, 12–13, 14, 22–23
    relationship to educational television, x
    Wisconsin Idea and, 2
    See also educational broadcasting; educational television; state radio network; Wisconsin

Public Radio; Wisconsin
Public Television
educational television
challenges to, 72–73, 74–76
frequencies reserved for, 66, 71
funding for, 66
relationship to educational
radio, x
social role of, 65
See also educational broadcasting;
educational radio; state
television network; WHA-TV
Madison; Wisconsin Public
Radio; Wisconsin Public
Television
Educational Television and Radio
Center (ETRC), 66, 70–71
Elkins, Stephanie, *186*
Ellison, Bob, 56
Engel, Harold, *82*
clear channel advocate, 26–27, 48
defense of not-for-profit
broadcasting, 29
educational broadcasting justified
by, 41, 106
noncredit college courses
proposed by, 35
philosophy of educational
broadcasting, 95
political forums proposed by,
42–44
state educational radio network
advocate, 52, 54–55
television advocate, 65–66,
72–74, 76
UW Extension advocate, 81
WHA, career at, 62, 32–33,
45–46, 82
"Enjoying Your Leisure" (program), 36
entertainment in educational
broadcasting, 19–20, 38, 44–45,
138
Estes, Bill, *147*
Etcetera, 63
ETRC. *See* Educational Television and
Radio Center (ETRC)
evaluation of educational radio, 40–41
Evjue, William, 31

Ewbank, Henry, *34*
chairmanships, 22, 49
open discussion supported by, 22,
50
philosophy of education vs.
entertainment, 45, 53
radio forums proposed by, 42
WHA programming leadership,
25–26, 28, 33, 40

faculty participation in programming
9XM, 14
agricultural journalism, 33
controversial topics, 58
lecture series, 17–18
music, 60
UW Extension, 84–85
WHA, 22
WHA-TV, 68
"Farm Life and Living" (program), 36
farm programs, 15, 33
Farm Show, 33, 59, 137
Farris, Charley, *61*
Federal Communication Commission
(FCC), 48, 66, 108, 156
Federal Radio Commission (FRC)
clear channel applications to,
26–27
state station applications to,
28–29
Feldman, Michael, 139–140, *140*
Fellman, David, 1–2
Feraca, Jean, *137*, 138, 177, 186–187
Ferrett, Rob, 184, *185*
first broadcast, 11
Fleming, Jim, *129*
announcer, 134
host of *To the Best of Our
Knowledge*, 175
on-air fund-raising, 142
program director, 177
Fleming, Robben, 80
Fletemeyer, Tom, 165, 182
FM frequencies reserved for non-
commercial educational use, 48
Follow the Flag, 156
Ford, Henry, 66
Ford Foundation, 66, 72, 73, 121

Frank, Boris, 77, 90
Frank, Glenn
  belief that radio shapes
    democracy, 23, 42
  philosophy of educational
    broadcasting, 44, 95, 138
  president of University of
    Wisconsin, 20–23
  testimony to Federal Radio
    Commission, 28
  WHA management appointed by,
    25
Frank, Lee, 110
Fred, Edwin, 52
frequencies reserved for educational
  programming, 48, 66, 71
Freyberg, Frederica, 167
Friendly Giant, The, 68–71, *70*
Friends of WHA-TV, 109, 115–119,
  165, 166
Friends of Wisconsin Public
  Television, 166
funding
  federal support of adult
    educational television, 92,
    94
  federal support of public
    broadcasting, 76, 99, 142
  Ford Foundation, 66, 72, 73, 101,
    105
  legislature and the television
    network, 73–76
  National Endowment for the Arts,
    123, 125
  National Science Foundation, 156
  on-air fund-raising, 115, *117*, 118,
    141–142, *143*
  public vs. private, 5
  public subsidy vs. individual
    subscriptions, 190–191
  Rockefeller Foundation, 40–41
  special-interest groups, 161
  sponsorships, 166
  state support of single station, 28
  state support of state FM network,
    53, 57
  state tax supported broadcasting,
    8, 120, 150, 189

university as funding source,
  21–22, 28, 189
US Office of Education, 90–92

Gard, Robert, 68
Gasser, Sheryl, *185*
General Extension division, 4, 16, 79,
  81
George, Gary, 166
GET REAL!, 156
Ghandi, Mahatma, 23
GooN Park (600 N. Park), 67–68, *67*
Gordon, Edgar, 18, 33, *34*, *38*, *40*, 56
governmental and political
  programming. *See* political
  programming
Governor's Report, 58
Grant, Joe, 128–130
Great War, 7
Groppi, James, 59, 88

"Half Hour of Useful Information, A"
  (program), 33
Hanna, Don, 180–181
Hanson, Malcolm, 11, *13*, 14, 34
Harley, William, 50, 76, 77
Harrington, Fred Harvey, 79–80, 108,
  109, 120
Hazard, Aline, 33–35, 68, 136
Heil, Julius, 48
Here and Now, 167, 168
Here on Earth, Radio without Borders,
  186, 187
Hess, Ronnie, 128, *129*
High Noon, 139
home economics programs, 33–35,
  36, 156
Homemakers' Program, 33–35, 68
hometown stories, 168–172
Homme, Bob, 68–71, *70*
Hopkins, Andrew, 14, 15, 22, 33, *34*,
  59
how-to programming, 166–167
Huffman, Jerry, 159, 161, *164*
humanities programming, 36, 177

Ideas Network
  listenership, 178

philosophy of, 175–176
programming of, 177–181, 183–188
WHA and, 176
Wisconsin Idea and, 176–181,
    183–185, 186–188, 193
See also Wisconsin Public Radio
Inner Core, The, 87–89
Internet, 185–188
interview programs, 169, 175
interviewing styles, 159–160
Iverson, Dave, 164
    Uncommon Places host, 114
    We the People, Wisconsin anchor,
        160, 163
    Weekend anchor, 158–159, 163
    Wisconsin Collaborative Project
        leader, 156
    Wisconsin Magazine host, 113

Jansky, C. M., 49
jazz music, 62
    See also music programming
Johnson, Ralph, 87, 127, 128, 130
Johnson, Ric, 69
Joseph, Matt, 138, 181
Journeys in Musicland, 38, 56

Kaleidoscope, 63
Kane, Charles Monroe, 187
Kentzler, Claire, 130, 141
Knight, Byron, 154
    commercial television
        partnership, 156–158
    digital television advocate,
        163–165
    partnerships with organizations
        for programming, 155–158
    program director of ECB, 112
    program director of ECB and
        WHA-TV, 153–154, 166
Knowles, Warren, 86, 88, 107, 108
Kohler, Walter, 73, 75
Kopit, Arthur, 123, 123, 124
Korean War documentaries, 169
Krulevitch, Walter, 51–52, 53

La Follette, Bob, 2
La Follette, Phillip, 28, 31, 42, 44

Lamb, Luke, 145, 151, 152
language instruction programs, 68
Lawson, Richard, 152, 153
League of Wisconsin (Commercial)
    Radio Stations, 53
legislation
    digital broadcasting mandate, 163
    ECB and, 190
    hearings, 181
    progressive movement and, 3
    Public Broadcasting Act of 1967, x,
        95, 120
    state radio network established by,
        52, 53
    state television network
        established by, 107, 163
    Wisconsin Act 9, 164
Legislative Audit Bureau, 145, 181
Legislative Reference Bureau, 3–4
Let's Draw, 39–40, 71
Let's Sing, 56
liberal arts educational programming,
    36, 59–60
liberalism in education, 58
licensing, 20
Lighty, William, 21, 34
    adult education advocate, 14–15, 17
    philosophy of educational
        broadcasting, 16, 30, 52, 95
    program director of 9XM, 14,
        18–19, 22
Listening, 124
localism and local programming,
    100–101, 114, 127, 144–145, 150,
    168–169
Loew, Patty, 164
Loftus, Tom, 159
Louris, Gary, 186
Lutz, Dick, 100–102, 103, 105–106
Lynch, Peg, 123
Lynde and Harry Bradley Foundation,
    170
LZ Lambeau, 170, 171

Macken, Mary, 130
Madden, John, 123, 124–125
Madison Vote-in, 101
Mankiewicz, Frank, 125

marches (music), 60
   See also music programming
market reports
   state department of agriculture, 15
   WHA, 33
McCarthy, Charles, 2–5, *3*, 14, 16, 36,
   79
McCarty, Harold B., *27*, *82*
   Dreyfus, conflict with, 77–80
   education vs. public service, 56
   philosophy of educational
      broadcasting, 37–38, 95
   program philosophy of, 55–56,
      57, 78
   public broadcasting theory, 58–59
   relationship to McNeil, 81, 83
   state FM stations supported by, 55
   television as an educational tool,
      67
   television pioneer, 65, 66, 76
   UW Extension and, 81
   WHA manager, 25–27, 30–33,
      39–53, 62, 79, 82
McKinley, Kevin, 183
McMahon, Nick, *172*
McMunn, Travis, *172*
McNeel, Wakelin, 38–39, 56
McNeil, Don
   head of UW Extension, 79, 84,
      85–87
   The Inner Core, 87, 89
   relationship to McCarty, 81, 83
Meet Our American Allies, 50
Meiller, Larry, 137, *137*, 138, 177
Minnesota Public Radio, 125, 179
Miskowski, Jon, 169–170, 171, 173
Mitchell, Jack, *132*, *147*
   cooperation with ECB staff,
      132–133, 143–148
   director of radio, ECB and UW
      Extension, 148–150
   director, Wisconsin Public Radio,
      ix, 148–150, 176, 180, 181
   idea for complementary
      programming for radio,
      132–133
   view of public radio's future,
      174–177

WHA manager, 1, 128, 131–148
   See also Ideas Network; Wisconsin
      Public Radio
Moe, Tony, 139
Monschein, Libby, 59
Monschein, Robert, 60
Moore, Andy, 158–159, 162–163, 167
Morning Edition, 135, 148
Music in Context, 60, 63
music programming
   appreciation of, 36, 38
   classical, 1, 16, 18, 60, 112
   concerts, 60
   jazz, 62
   marches (music), 60
   on Wisconsin Public Radio,
      133–134

National Center for Audio
   Experimentation, 121
National Center for Media
   Engagement, 170–171
National Endowment for the Arts
   (NEA), 123, 125
National Public Radio (NPR)
   audience, 193
   establishment of, 95, 99
   programming, 121
   radio dramas and, 125
National Science Foundation, 156
nature studies programming, 38, 56
Nelson, Gaylord, 58, 101
New Tech Times, The, 118
news programs
   Views of the News, 59
   Weekend, 158–160, 162–163
   on WHA-AM, 1
   wire services as sources, 59
   on Wisconsin Public Radio,
      135–140, 176
   World War II reports, 50
newspapers, 6
noncommercial radio. *See* educational
   radio; public broadcasting; state
   radio network; Wisconsin Public
   Radio
noncommercial television. *See*
   educational television; public

broadcasting; state television
network; Wisconsin Public
Television
Norton, Paul, *147*
    executive director of ECB, 133,
        139, 142–144, 147
    statewide dual service plan,
        support for, 144–145
    WHA-TV and ECB cooperation
        supported by, 152–153
    Wisconsin Public Radio
        partnership supported by,
        147–148
notmuch.com, 186
Nova, 114
NPR. *See* National Public Radio
NPR News and Classical Network,
    177, 178, 193

Obey, David, 107
Ohio State University, 38, 122
Ohst, Ken, *61*
    death of, 130
    The Friendly Giant, 70–71
    music programs, 60, 62, 134
    sports programs, 68
on-demand programming, 185–188,
    190
Out of Context, 128

Packard, Jim, 140, *140*
Party Time, 42
Paulson, Steve, *137*, 175
PBS. *See* Public Broadcasting Service
pledge drives, 142, 173, 178
political debates, 42–43, 159–163
Political Education Forum, The, 42–43,
    44
political programming, 43, 58,
    102–103, 158–162
Pollyanna Views the News, 105
Powell, John, 128
Pretty Soon Runs Out, 88
Price, John, 117–119
print media, 5–6
programming
    on aging, 89–90

agricultural, 15, 27–28, 33, 36,
    45, 59
art, 36, 39–40
awards for, 71, 88, 112, 116, 124,
    187
call-in, 135–138
car advice, 138, 181
career education, 90–94
children, 37–41, *39*, 68–71, 156,
    166–167
classical music, 1, 16, 18, 60, 112
controversial, 50, 51, 58,
    102–103, 116, 128–129, 153
cultural, 16, 36, 49, 68, 168
entertainment vs. educational,
    19–20, 38, 44–45, 138
home economics, 33–35, 36, 156
how-to, 166–167
humanities, 36, 177
inner city issues, 86–89
interview, 169, 175
language instruction, 68
liberal arts, 36, 59–60
local, 100–101, 114, 127, 144–145,
    150, 168–169
music, 16, 18, 60, 62, 112, 133–134
nature studies, 38, 56
news and informational, 1, 15, 50,
    59, 135–140, 158–163, 176
on-demand, 185–188, 190
political, 43, 58, 102–103, 158–162
public affairs, 57, 158–163
public service, 20, 56–57, 59, 107
recorded, 56–57, 126, 167
school, 37–41, *39*, 56, 71
social and natural sciences, 36
sports, 18, 45, 54, 68, 111
university courses, 36–37, 37
UW Extension, 83–86, 89–90
weather, 15, 33, 61–62, 127
youth oriented, 37, 38, 184
See also documentaries; state
    stations; state television
    network; University of
    Wisconsin Extension;
    Wisconsin College of the Air;
    Wisconsin Public Radio;

Wisconsin Public Television;
and *names of individual
programs*
progressive reform movement
legislation and, 3
McCarthyism impact on, 72
newspapers' and magazines' role
in, 5–6
Wisconsin Idea and, 2–3, 5
World War I impact on, 7
public affairs programming, 57,
158–163
public broadcasting
definition of, 99–100, 114–115,
116, 135
educational broadcasting vs., 99
funding for, 140, 151, 189–191
informative and entertaining,
192–193
service to democracy, ix
university outreach goals, 192
Wisconsin Idea, foundation of,
ix–x, 12–13, 14, 52, 176,
189–194
Public Broadcasting Act of 1967, x,
95, 120
Public Broadcasting Service (PBS), 95,
99, 114, 155, 166
Public Expenditure Survey of
Wisconsin, 74
public radio
audience, 128
funding for, 121
National Public Radio and, 121
programming, centralized, 121,
176
Public Broadcasting Act of 1967,
99
relationship to public television, x
university students and, 128
See also Educational
Communications Board;
public broadcasting; state
radio network; state stations;
WHA; Wisconsin Public Radio
public radio vs. commercial radio, 2,
28, 30, 31

public service programming, 20,
56–57, 59, 107
public support. *See* audiences
public television
audience reaction, 103–105
competition from cable and
satellite alternatives, 151
funding for, 101, 105
legislation and integration of ECB
and WHA-TV, 152
legislators' reaction, 106
localism served by, 100–101
programming of, 101–103, 106
Public Broadcasting Act of 1967,
99
viewership, 151
See also public broadcasting; state
television network; WHA-TV
Madison; Wisconsin Public
Television
Purcell, Gene, 166

Quiz the Professor, 68

radio
adult education, 15
beginnings, 11–14
in classroom, 37–38, 126
dramas, 122–126
forums, 42
relation to education, 7–8, 28,
30, 35
university boundaries extended
by, 7
Wisconsin Idea and, 7
See also programming; state
radio network; state stations;
Wisconsin Public Radio; and
names of individual programs
Radio Hall, 32–33, *32*, 67
radio stations, 7, 13–14, 57, 59, 120.
See also state stations; Wisconsin
Public Radio; and *names of
individual stations*
Ranger Mac, 71
Ranger Mac (Wakelin McNeel),
38–39, 56

Ratner, Mindy, 128, *129*
Reagan administration, 151, 190
recording of lectures, 56–57, 167
Red Gym broadcast site, *19*
Rehm, Dana, 181–182
Rennebohm, Oscar, 52, 53, 55
Restructuring Public Broadcasting
    and Funding Digital Television
    Transition Committee, 164–166
Reynolds, John, 58
Rhythm and Games, 38, 71
Rieland, Allen, 185
Roberts, Cliff, 60, 127, 130, 134
Robertson, Jim
    philosophy of control of public
        broadcasting, 88
    priorities for state radio network
        and WHA-TV, 83
    proposal for state television
        network, 106–107
    resignation of, 108
Rockefeller Foundation, 40–41, 46
Roosevelt, Theodore, 4
Rueckert, Veronica, 184, *185*
rural audiences, 15, 33–35, 90–94
Rural Family Development (RFD),
    90–92
Russell, Robert, 142
Ryan, Shelley, 155, *157*

satellite radio, 174, 175
Schmedeman, Albert, 31
Schmidt, Karl
    director of radio drama center,
        122–126
    director of radio at UW Extension,
        83, 89, 121
    philosophy of on-air fund-raising,
        142
Schnirring, Greg, 178, 180, 181, 182
School of the Air, 37–39, 56, 71, 110
school programming, 37–41, *39*, 56, 71
Schwalbach, James, 39–40
Science Hall, 7, 11, 12, *12*
science programs, 36
Search for the Violin, 114
Selling of America, The, 116
Sewing with Nancy, 156

Siemering, William, *63*, 193
Siepmann, Charles, 46–47
Silber, Glenn, 112
Simply Folk, 134
SiriusXM, 175
Six30, 102–106, 108
"Social Problems Today" (program),
    36
Soglin, Paul, 102–103
sports programs, 18, 45, 54, 68, 111
Sprangers, Lynn, *164*
State Building Commission, 166
State Radio Council
    creation of, 49
    ECB formed from, 107
    FM statewide network plan, 50
    goals of, 55
    independence from university
        control, 50–51
    legislation establishes, 52
    opposition to, 55
    policy regarding commercial
        stations, 54
    relation to UW system, 50–52
    structure and composition of, 49,
        52
    television license awarded to, 66
    Wisconsin Idea and, 49
state radio network
    budgets, 72
    ECB and, 126–127
    expansion of, 52
    funding for, 51, 107
    independence from university, 50
    legislation authorizing, 52
    programming of, 56–62, 63–64,
        88, 127
    reactions to, 52–55
    Wisconsin Idea and, 52
    See also Education
        Communications Board
        (ECB); educational radio;
        public radio; Wisconsin Public
        Radio
state stations
    benefit of, 28
    competition from commercial
        stations, 30–31

evaluation of educational
    programming, 40–41
expansion of, 59
funding for, 39
military training programs, 50
programming and service to
    public, 28–29, 33–40, 42–46,
    50, 56
shared programming, 30
war programming, 50
Wisconsin Educational Radio
    Network, 126
WLBL Stevens Point, 27–28, 33
See also WHA; Wisconsin College
    of the Air; Wisconsin Public
    Radio; Wisconsin School of the
    Air; and names of individual
    programs
state television network
budgets, 72
Channel 10 Milwaukee, 73, 76, 88
Channel 21 Madison, 71, 73, 76
controlled by, 106, 108
debate over funding, 73–76
as educational tool, 78
funding for, 66, 77, 106, 107, 127
legislation and, 107
programming of, 77, 107, 110,
    112–114
ratings, 77
referendum for, 73–76
Robertson's proposal for, 106–107
value of, 65, 73
Wisconsin Idea and, 78–79
See also Educational
    Communications Board
    (ECB); public broadcasting;
    public television; Wisconsin
    Public Television
statewide education, 7, 79
station identifications, 62, 127
Steinbach, James
commercial television
    partnership, 156–158
director, Wisconsin Public
    Television, 153–154, 160, 168
LZ Lambeau and, 170
Sterling Hall, 12, 18, 21, 31, 32

Steve, Fannie, 38–39, 39, 71
Strainchamps, Ann, 175, 187
student participation in
    programming, 63
9XM, 14
agricultural journalism, 33
controversial topics, 51, 116, 128
WHA, 25, 63
WPR, 149
Swanton, Milo, 73, 75

tax-supported broadcasting, ix, 8,
    120, 150, 189
television in the classroom, 71, 106,
    109
television stations, heavily funded,
    121
television studios, 67–68, 160
Terry, Earle, 13, 21, 34
death of, 25
development of wireless
    transmitter for radio, 7, 11–16
philosophy of educational
    broadcasting, 95, 192
social vision of, 14
Wisconsin Idea and, 7, 12–13, 14
Thompson, Tommy, 159, 162,
    164–166
Tiano, Tony, 92, 94
Time of Our Lives, The, 89
To the Best of Our Knowledge, 175, 187
Tongues Untied, 153
Tryout TV, 116
Tungekar, Rehman, 187

Uncommon Places, 114
underwriting, 182, 183
university courses on radio, 36–37,
    37, 51, 56, 63
University Forum, 57
University of Wisconsin
    600 N. Park (GooN Park), 67–68,
        67
    agricultural extension, 4
    antagonism with ECB, 152
    athletic events broadcast, 18, 54
    budget effects on public
        broadcasting, 189

radio station, 11, 144–150
relationship to State Radio
    Council, 50–52
Science Hall, 7, 11, 12, *12*
statewide education as goal, 7
Sterling Hall, 12, 18, *21*, 31, 32
Vilas Hall, 13, 68, *91*, 92, 110, 126
WHA-AM and, 126
Wisconsin Idea as mission of, 4, 189
University of Wisconsin Extension
agricultural extension, 4, 79, 81
Collins, radio programming
    manager, 84
ECB, tensions with, 126, 147–148
faculty of, 81, 84–85, 89, 94–95
faculty vs. educational
    broadcasters, 84
General Extension, 4, 16, 79, 81
head of, 79
inner core project, 87–89
McNeil's support for, 84
Mitchell, director of radio, 148–150
purpose of, 79, 81
radio programming for general
    and student audiences, 83,
    84–86, 89–90, 144
Robertson's priorities for, 83
Schmidt, director of radio at UW
    Extension, 83, 89, 121
television programming advisory
    committee, 85, 95
television programming for
    specialized audiences, 89–95
Wisconsin Idea and, 79, 83, 95
University of Wisconsin Roundtable, 58
University Place, 167
University Radio Committee, 15, 22,
    34, 49, 51
US Office of Education, 40, 90
UW Extension. *See* University of
    Wisconsin Extension
UW Roundtable Student Forum, 51

vacuum tubes, 13
Van Hise, Charles, 2, 4, *5*, 6, 14, 52
Vel Phillips, 171–172
veterans' stories, 169–170
Vietnam era documentaries, 169–170

Views of the News, 59
Vilas Hall, 13, 68, *91*, 92, 110, 126
"Vilas Talk," 176
vocational education, 36–37
Voegeli, Don, 121, *122*
Voegeli, Tom, 124–125
Vogelman, Roy, 58, 59, 61, 68, 130

war history programs, 169–170
Wardenburg, Fred, 88
We the People, Wisconsin, 158, 160–163
weather programming, 15, 33, 61–62,
    127
Weather Roundup, The, 61–62
Weekend, 158–160, *160*, 162–163, *164*
Weller, Jan, *137*, 146
WERN, 126, 176, 178. *See also*
    WHA-FM Madison
Weston, Tom, 111–112
WGBW Green Bay, 148–149
WGN Chicago, 140
WHA
    9XM replaced by, 14
    athletic events broadcast on, 18
    clear channel application, 27, 44
    College of the Air, 35–36, *37*, 51,
        56, 63
    funding for, 121, 141–142
    management of, 25
    McCarty, manager of, 25–27,
        30–33, 39–53, 62, 79, 82
    Mitchell, manager, 1, 128, 131–148
    programming of, 36, 56–62,
        63–64, 121–126, 127–130
    relationship with UW Extension,
        95
    School of the Air, 37–39, 56, 71, 110
    staff camaraderie, 63–64
    studios for, 12, 18, 31–33, 92
    UW system support for, 22
    See also state stations, Wisconsin
        Idea
WHA Madison (AM station)
    Ideas Network station, 178
    NPR and, 135, 140
    programming of, 132, 135–140, 176
    See also state stations; Wisconsin
        Public Radio

WHA-FM Madison, 52, 118–119, 126
WHA-TV Madison, *67*
    beginnings, 66–68
    community focus, 116
    control of, 108–109
    ECB and, 108–110, 112, 113–114,
        152
    educational mission, 112
    funding for, 110, 112, 114
    Knight, program director,
        153–154, 165–166
    Lutz's management of, 101–105
    Lutz's philosophy of television, 101
    on-air fund-raising, 115, *117*, 118
    programming of, 68–71, 77, 88,
        101–105, 110–114, 116, 118
    ratings, 111
    reactions to, 102–105
    studio, 110
    Wisconsin Idea and, 78–79
    See also Wisconsin Public
        Television
WHAD Delafield, 52, 146, 179
Whad Ya Know?, 140, 185–186
Whaley, K. P., *185*
White, Maurey, 59
WIBA Madison, 27, 30, 31
Willis, Jack, 88
Winds of Change, 114
wireless transmitting device, 7
WIS Stevens Point, 28
Wisconsin Act 9, 164
Wisconsin Alert, 50
Wisconsin Channel, 167–169, 171–173
Wisconsin Citizens Committee for
    Educational Television, 73, 75
Wisconsin Collaborative Project, 156
Wisconsin College of the Air, 35–37, *37*,
    51, 56, 63
Wisconsin Committee on State-Owned,
    Tax-Supported Television, 74
Wisconsin department of agriculture
    (state agency)
    creation of WLBL, 27–28
    sports broadcasts funded by, 45
Wisconsin Educational Radio Network
    (WERN)
    control of, 132

programming of, 126, 127, 128,
    132–134, 176, 178
ratings book, 178
state stations and, 126
WHA-FM renamed, 126, 132
See also Wisconsin Public Radio
Wisconsin from the Air, 172–173
Wisconsin Gardner, The, 155
Wisconsin Historical Society, 168,
    169
Wisconsin Hometown Stories, 168–169
Wisconsin Idea
    as basis for public broadcasting,
        12–13, 14, 52, 176, 189–190,
        192, 193
    funding of, 8, 190
    Ideas Network connection,
        176–181, 183–185, 186–188,
        193
    objective of Wisconsin Public
        Radio and Wisconsin Public
        Television, 189, 194
    philosophy of, x, 2–4, 190–192
    progressive movement and, 2, 5, 7
    radio's role in 7, 23–24
    staff characterized by, 185
    university and 4, 7, 189, 191
    UW Extension and, 79
    UW system and, 2, 4, 7, 189, 191
Wisconsin Idea, The (book)
    (McCarthy), 2–3, 5
Wisconsin Legislative Forum, 43
Wisconsin Life, 168
Wisconsin Magazine, 113–114
Wisconsin Public Radio
    budget, 189
    call-in programming, 135–138
    creation of, 133
    digital broadcast technology, 177,
        182
    directors for, 181–183
    dual program streams, 132–133,
        143–144, 175–176
    ECB-UW Extension conflict,
        147–148
    equipment upgrades, 182
    expansion of, 144–150, 182
    funding for, 141, *143*, 150, 182, 183

Ideas Network and, 176–177, 178, 193
Internet, 184–188
Legislative Audit Bureau endorsement, 145
Mitchell, director, ix, 148–150, 176, 180–181
Mitchell's view of public radio's future, 174–177
music programming, 133–134, 178, 180, 186
national reputation, 181
news programming, 135–136, 176, 186, 188
public affairs programming, 158–163, 183
satellite broadcast, effect on programming, 174–175
service to state, 188, 190
staffing, 185
To the Best of Our Knowledge, 175, 187
UW system opposition, 144–146
Wisconsin Idea and, 188, 189, 194
wpr.org, 186
youth-oriented programming, 184
See also Educational Communications Board (ECB); University of Wisconsin; University of Wisconsin Extension; WHA
Wisconsin Public Radio Association, 142
Wisconsin Public Television
budget, 153, 189
commercial television and, 156
creation of, 151–152
digital broadcast technology, 163, 166
funding for, 155–157, 163–166, 170
local programming, 153, 155, 168–169
merger of ECB and UW Extension operations, 153
partnerships of, 155–157, 161, 166, 168–169
programming of, 154, 155–163, 166–173, 193

PBS and, 155, 166–167
public perception of, 154
service to state, 153, 154, 172, 190
Steinbach, director, 153–154, 160, 168
UW Extension and, 155
Wisconsin Channel, 167–169
Wisconsin communities as program subjects, 168–169
Wisconsin Idea and, 172–173, 189, 194
See also Educational Communications Board (ECB); University of Wisconsin; University of Wisconsin Extension; WHA-TV Madison
Wisconsin School of the Air, 37–39, 56, 71, 110
Wisconsin State Radio Network. *See* state radio network
Wisconsin Veterans Museum, 169
Wisconsin Week, 114
WLBL Stevens Point, 27–28, 33
WMAQ Chicago, 44, 48
women in broadcasting
Homemakers' Program, 33–35
managers, 130, 181
on-air personalities, 128, 134, 148, 155–156, 177–178, 183–184, 186
writers, 56, 123
Wonderful World of Nature, The, 56
Wood, Jim, 161
"World about You, The" (program), 36
World War I, 7
World War II, 50, 169
wpr.org, 186, 188
WTMJ Milwaukee, 27, 30
WUWM Milwaukee, 144, 145, 146, 179
WYMS Milwaukee, 146

XM radio, 174

"You and Your Home" (program), 36
youth-oriented programming, 37, 38, 184

ABOUT THE AUTHOR

Jack Mitchell, PhD, led Wisconsin Public Radio for twenty-one years, from 1976 till 1997, initiating the transition from the Wisconsin Educational Radio Network into Wisconsin Public Radio. On a national level, Mitchell was the first employee of National Public Radio, where he was instrumental in developing the groundbreaking newsmagazine *All Things Considered*. During his years as the program's first producer and executive producer, the program won both the Peabody Award and the DuPont Award. Mitchell was elected to an unprecedented four terms on the National Public Radio

UW SCHOOL OF JOURNALISM AND MASS COMMUNICATION

Board of Directors, including three years as chair. He received the two highest honors in public radio, the Corporation for Public Broadcasting's Edward R. Murrow Award and the Edward Elson National Public Radio Distinguished Service Award. Mitchell joined faculty of the UW–Madison School of Journalism and Mass Communication in 1998. He is the author of *Listener Supported: The Culture and History of Public Radio*.